LINEAR ALGEBRA AS
AN INTRODUCTION TO
ABSTRACT MATHEMATICS

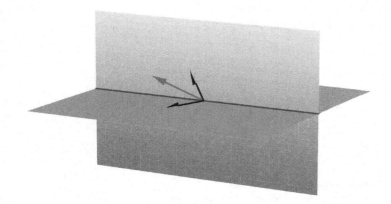

LINEAR ALGEBRA AS
AN INTRODUCTION TO
ABSTRACT MATHEMATICS

Isaiah Lankham
California State University, East Bay, USA

Bruno Nachtergaele
University of California, Davis, USA

Anne Schilling
University of California, Davis, USA

World Scientific

NEW JERSEY · LONDON · SINGAPORE · BEIJING · SHANGHAI · HONG KONG · TAIPEI · CHENNAI · TOKYO

Published by

World Scientific Publishing Co. Pte. Ltd.

5 Toh Tuck Link, Singapore 596224

USA office: 27 Warren Street, Suite 401-402, Hackensack, NJ 07601

UK office: 57 Shelton Street, Covent Garden, London WC2H 9HE

Library of Congress Cataloging-in-Publication Data

Names: Nachtergaele, Bruno. | Schilling, Anne (Mathematician) | Lankham, Isaiah.
Title: Linear algebra as an introduction to abstract mathematics / by Bruno Nachtergaele (UC Davis),
    Anne Schilling (UC Davis), Isaiah Lankham (California State University, USA).
Description: New Jersey : World Scientific, 2016. | Includes bibliographical references and index.
Identifiers: LCCN 2015040257 | ISBN 9789814730358 (hardcover : alk. paper) |
    ISBN 9789814723770 (softcover : alk. paper)
Subjects: LCSH: Algebras, Linear.
Classification: LCC QA184 .N33 2016 | DDC 512/.5--dc23
LC record available at http://lccn.loc.gov/2015040257

British Library Cataloguing-in-Publication Data

A catalogue record for this book is available from the British Library.

Printed in Singapore

Preface

"Linear Algebra - As an Introduction to Abstract Mathematics" is an introductory textbook designed for undergraduate mathematics majors and other students who do not shy away from an appropriate level of abstraction. In fact, we aim to introduce abstract mathematics and proofs in the setting of linear algebra to students for whom this may be the first step toward advanced mathematics. Typically, such a student will have taken calculus, though the only prerequisite is suitable mathematical maturity. The purpose of this book is to bridge the gap between more conceptual and computational oriented lower division undergraduate classes and more abstract oriented upper division classes.

The book begins with systems of linear equations and complex numbers, then relates these to the abstract notion of linear maps on finite-dimensional vector spaces, and covers diagonalization, eigenspaces, determinants, and the spectral theorem. Each chapter concludes with both proof-writing and computational exercises.

We wish to thank our many undergraduate students who took MAT67 at UC Davis in the past several years and our colleagues who taught from our lecture notes that eventually became this book. Their comments on earlier drafts were invaluable. This book is dedicated to them and all future students and teachers who use it.

I. Lankham

B. Nachtergaele

A. Schilling

California, October 2015

Contents

Chapter 1

What is Linear Algebra?

1.1 Introduction

This book aims to bridge the gap between the mainly computation-oriented lower division undergraduate classes and the abstract mathematics encountered in more advanced mathematics courses. The goal of this book is threefold:

(1) You will learn **Linear Algebra**, which is one of the most widely used mathematical theories around. Linear Algebra finds applications in virtually every area of mathematics, including multivariate calculus, differential equations, and probability theory. It is also widely applied in fields like physics, chemistry, economics, psychology, and engineering. You are even relying on methods from Linear Algebra every time you use an internet search like Google, the Global Positioning System (GPS), or a cellphone.

(2) You will acquire **computational skills** to solve linear systems of equations, perform operations on matrices, calculate eigenvalues, and find determinants of matrices.

(3) In the setting of Linear Algebra, you will be introduced to **abstraction**. As the theory of Linear Algebra is developed, you will learn how to make and use definitions and how to write proofs.

The exercises for each Chapter are divided into more computation-oriented exercises and exercises that focus on proof-writing.

1.2 What is Linear Algebra?

Linear Algebra is the branch of mathematics aimed at solving systems of linear equations with a finite number of unknowns. In particular, one would like to obtain answers to the following questions:

- **Characterization of solutions:** Are there solutions to a given system of linear equations? How many solutions are there?
- **Finding solutions:** How does the solution set look? What are the solutions?

Linear Algebra is a systematic theory regarding the solutions of systems of linear equations.

Example 1.2.1. Let us take the following system of two linear equations in the two unknowns x_1 and x_2:

$$\left. \begin{array}{r} 2x_1 + x_2 = 0 \\ x_1 - x_2 = 1 \end{array} \right\}.$$

This system has a **unique solution** for $x_1, x_2 \in \mathbb{R}$, namely $x_1 = \frac{1}{3}$ and $x_2 = -\frac{2}{3}$.

The solution can be found in several different ways. One approach is to first solve for one of the unknowns in one of the equations and then to substitute the result into the other equation. Here, for example, we might solve to obtain

$$x_1 = 1 + x_2$$

from the second equation. Then, substituting this in place of x_1 in the first equation, we have

$$2(1 + x_2) + x_2 = 0.$$

From this, $x_2 = -2/3$. Then, by further substitution,

$$x_1 = 1 + \left(-\frac{2}{3} \right) = \frac{1}{3}.$$

Alternatively, we can take a more systematic approach in eliminating variables. Here, for example, we can subtract 2 times the second equation from the first equation in order to obtain $3x_2 = -2$. It is then immediate that $x_2 = -\frac{2}{3}$ and, by substituting this value for x_2 in the first equation, that $x_1 = \frac{1}{3}$.

Example 1.2.2. Take the following system of two linear equations in the two unknowns x_1 and x_2:

$$\left. \begin{array}{r} x_1 + x_2 = 1 \\ 2x_1 + 2x_2 = 1 \end{array} \right\}.$$

We can eliminate variables by adding -2 times the first equation to the second equation, which results in $0 = -1$. This is obviously a contradiction, and hence this system of equations has **no solution**.

Example 1.2.3. Let us take the following system of one linear equation in the two unknowns x_1 and x_2:

$$x_1 - 3x_2 = 0.$$

In this case, there are **infinitely many** solutions given by the set $\{x_2 = \frac{1}{3}x_1 \mid x_1 \in \mathbb{R}\}$. You can think of this solution set as a line in the Euclidean plane $\mathbb{R}^2$:

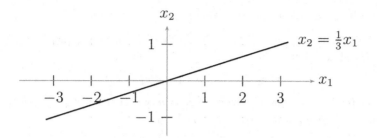

In general, a **system of m linear equations in n unknowns** $x_1, x_2, \ldots, x_n$ is a collection of equations of the form

$$\left.\begin{array}{c} a_{11}x_1 + a_{12}x_2 + \cdots + a_{1n}x_n = b_1 \\ a_{21}x_1 + a_{22}x_2 + \cdots + a_{2n}x_n = b_2 \\ \vdots \qquad \vdots \\ a_{m1}x_1 + a_{m2}x_2 + \cdots + a_{mn}x_n = b_m \end{array}\right\}, \tag{1.1}$$

where the a_{ij}'s are the coefficients (usually real or complex numbers) in front of the unknowns x_j, and the b_i's are also fixed real or complex numbers. A **solution** is a set of numbers $s_1, s_2, \ldots, s_n$ such that, substituting $x_1 = s_1, x_2 = s_2, \ldots, x_n = s_n$ for the unknowns, all of the equations in System (1.1) hold. Linear Algebra is a theory that concerns the solutions and the structure of solutions for linear equations. As we progress, you will see that there is a lot of subtlety in fully understanding the solutions for such equations.

1.3 Systems of linear equations

1.3.1 *Linear equations*

Before going on, let us reformulate the notion of a system of linear equations into the language of functions. This will also help us understand the adjective "linear" a bit better. A **function** f is a map

$$f : X \to Y \tag{1.2}$$

from a set X to a set Y. The set X is called the **domain** of the function, and the set Y is called the **target space** or **codomain** of the function. An **equation** is

$$f(x) = y, \tag{1.3}$$

where $x \in X$ and $y \in Y$. (If you are not familiar with the abstract notions of sets and functions, please consult Appendix B.)

Example 1.3.1. Let $f : \mathbb{R} \to \mathbb{R}$ be the function $f(x) = x^3 - x$. Then $f(x) = x^3 - x = 1$ is an equation. The domain and target space are both the set of real numbers $\mathbb{R}$ in this case.

In this setting, a system of equations is just another kind of equation.

Example 1.3.2. Let $X = Y = \mathbb{R}^2 = \mathbb{R} \times \mathbb{R}$ be the Cartesian product of the set of real numbers. Then define the function $f : \mathbb{R}^2 \to \mathbb{R}^2$ as

$$f(x_1, x_2) = (2x_1 + x_2, x_1 - x_2), \tag{1.4}$$

and set $y = (0, 1)$. Then the equation $f(x) = y$, where $x = (x_1, x_2) \in \mathbb{R}^2$, describes the system of linear equations of Example 1.2.1.

The next question we need to answer is, "What is a linear equation?". Building on the definition of an equation, a **linear equation** is any equation defined by a "linear" function f that is defined on a "linear" space (a.k.a. a **vector space** as defined in Section 4.1). We will elaborate on all of this in later chapters, but let us demonstrate the main features of a "linear" space in terms of the example $\mathbb{R}^2$. Take $x = (x_1, x_2), y = (y_1, y_2) \in \mathbb{R}^2$. There are two "linear" operations defined on $\mathbb{R}^2$, namely addition and scalar multiplication:

$$x + y := (x_1 + y_1, x_2 + y_2) \qquad \text{(vector addition)} \tag{1.5}$$

$$cx := (cx_1, cx_2) \qquad \text{(scalar multiplication).} \tag{1.6}$$

A "linear" function on $\mathbb{R}^2$ is then a function f that interacts with these operations in the following way:

$$f(cx) = cf(x) \tag{1.7}$$

$$f(x + y) = f(x) + f(y). \tag{1.8}$$

You should check for yourself that the function f in Example 1.3.2 has these two properties.

1.3.2   *Non-linear equations*

(Systems of) Linear equations are a very important class of (systems of) equations. We will develop techniques in this book that can be used to solve any systems of linear equations. Non-linear equations, on the other hand, are significantly harder to solve. An example is a **quadratic equation** such as

$$x^2 + x - 2 = 0, \tag{1.9}$$

which, for no completely obvious reason, has exactly two solutions $x = -2$ and $x = 1$. Contrast this with the equation

$$x^2 + x + 2 = 0, \tag{1.10}$$

which has no solutions within the set $\mathbb{R}$ of real numbers. Instead, it has two complex solutions $\frac{1}{2}(-1 \pm i\sqrt{7}) \in \mathbb{C}$, where $i = \sqrt{-1}$. (Complex numbers are discussed in more detail in Chapter 2.) In general, recall that the quadratic equation $x^2 + bx + c = 0$ has the two solutions

$$x = -\frac{b}{2} \pm \sqrt{\frac{b^2}{4} - c}.$$

1.3.3 *Linear transformations*

The set $\mathbb{R}^2$ can be viewed as the Euclidean plane. In this context, linear functions of the form $f : \mathbb{R}^2 \to \mathbb{R}$ or $f : \mathbb{R}^2 \to \mathbb{R}^2$ can be interpreted geometrically as "motions" in the plane and are called **linear transformations**.

Example 1.3.3. Recall the following linear system from Example 1.2.1:

$$\left. \begin{array}{r} 2x_1 + x_2 = 0 \\ x_1 - x_2 = 1 \end{array} \right\}.$$

Each equation can be interpreted as a straight line in the plane, with solutions (x_1, x_2) to the linear system given by the set of all points that simultaneously lie on both lines. In this case, the two lines meet in only one location, which corresponds to the unique solution to the linear system as illustrated in the following figure:

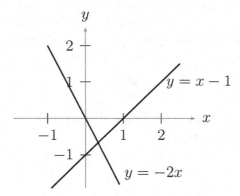

Example 1.3.4. The linear map $f(x_1, x_2) = (x_1, -x_2)$ describes the "motion" of reflecting a vector across the x-axis, as illustrated in the following figure:

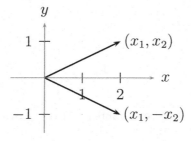

Example 1.3.5. The linear map $f(x_1, x_2) = (-x_2, x_1)$ describes the "motion" of rotating a vector by $90°$ counterclockwise, as illustrated in the following figure:

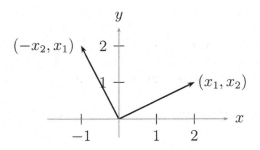

This example can easily be generalized to rotation by any arbitrary angle using Lemma 2.3.2. In particular, when points in $\mathbb{R}^2$ are viewed as complex numbers, then we can employ the so-called polar form for complex numbers in order to model the "motion" of rotation. (Cf. Proof-Writing Exercise 5 on page 20.)

1.3.4   *Applications of linear equations*

Linear equations pop up in many different contexts. For example, you can view the derivative $\frac{df}{dx}(x)$ of a differentiable function $f : \mathbb{R} \to \mathbb{R}$ as a linear approximation of f. This becomes apparent when you look at the Taylor series of the function $f(x)$ centered around the point $x = a$ (as seen in calculus):

$$f(x) = f(a) + \frac{df}{dx}(a)(x - a) + \cdots . \tag{1.11}$$

In particular, we can graph the linear part of the Taylor series versus the original function, as in the following figure:

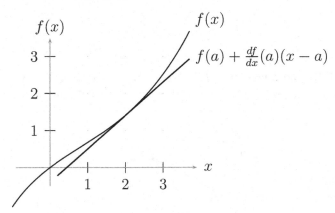

Since $f(a)$ and $\frac{df}{dx}(a)$ are merely real numbers, $f(a) + \frac{df}{dx}(a)(x-a)$ is a linear function in the single variable x.

Similarly, if $f : \mathbb{R}^n \to \mathbb{R}^m$ is a multivariate function, then one can still view the derivative of f as a form of a linear approximation for f (as seen in a multivariate calculus course).

What if there are infinitely many variables $x_1, x_2, \ldots$? In this case, the system of equations has the form

$$\left.\begin{array}{l} a_{11}x_1 + a_{12}x_2 + \cdots = y_1 \\ a_{21}x_1 + a_{22}x_2 + \cdots = y_2 \\ \qquad\qquad \cdots \end{array}\right\}.$$

Hence, the sums in each equation are infinite, and so we would have to deal with infinite series. This, in particular, means that questions of convergence arise, where convergence depends upon the infinite sequence $x = (x_1, x_2, \ldots)$ of variables. These questions will not occur in this course since we are only interested in finite systems of linear equations in a finite number of variables. Other subjects in which these questions do arise, though, include

- Differential equations;
- Fourier analysis;
- Real and complex analysis.

In algebra, Linear Algebra is also seen to arise in the study of symmetries, linear transformations, and Lie algebras.

Exercises for Chapter 1

Calculational Exercises

(1) Solve the following systems of linear equations and characterize their solution sets. (I.e., determine whether there is a unique solution, no solution, etc.) Also, write each system of linear equations as an equation for a single function $f : \mathbb{R}^n \to \mathbb{R}^m$ for appropriate choices of $m, n \in \mathbb{Z}_+$.

(a) System of 3 equations in the unknowns x, y, z, w:

$$\left.\begin{array}{l} x + 2y - 2z + 3w = 2 \\ 2x + 4y - 3z + 4w = 5 \\ 5x + 10y - 8z + 11w = 12 \end{array}\right\}.$$

(b) System of 4 equations in the unknowns x, y, z:

$$\left.\begin{array}{l} x + 2y - 3z = 4 \\ x + 3y + z = 11 \\ 2x + 5y - 4z = 13 \\ 2x + 6y + 2z = 22 \end{array}\right\}.$$

(c) System of 3 equations in the unknowns x, y, z:

$$\left.\begin{array}{l} x + 2y - 3z = -1 \\ 3x - y + 2z = 7 \\ 5x + 3y - 4z = 2 \end{array}\right\}.$$

(2) Find all pairs of real numbers x_1 and x_2 that satisfy the system of equations

$$x_1 + 3x_2 = 2, \tag{1.12}$$
$$x_1 - x_2 = 1. \tag{1.13}$$

Proof-Writing Exercises

(1) Let a, b, c, and d be real numbers, and consider the system of equations given by

$$ax_1 + bx_2 = 0, \tag{1.14}$$
$$cx_1 + dx_2 = 0. \tag{1.15}$$

Note that $x_1 = x_2 = 0$ is a solution for any choice of a, b, c, and d. Prove that if $ad - bc \neq 0$, then $x_1 = x_2 = 0$ is the only solution.

Chapter 2

Introduction to Complex Numbers

Let $\mathbb{R}$ denote the set of **real numbers**, which should be a familiar collection of numbers to anyone who has studied calculus. In this chapter, we use $\mathbb{R}$ to build the equally important set of so-called complex numbers.

2.1   Definition of complex numbers

We begin with the following definition.

Definition 2.1.1. The set of **complex numbers** $\mathbb{C}$ is defined as

$$\mathbb{C} = \{(x, y) \mid x, y \in \mathbb{R}\}.$$

Given a complex number $z = (x, y)$, we call $\text{Re}(z) = x$ the **real part** of z and $\text{Im}(z) = y$ the **imaginary part** of z.

In other words, we are defining a new collection of numbers z by taking every possible ordered pair (x, y) of real numbers $x, y \in \mathbb{R}$, and x is called the real part of the ordered pair (x, y) in order to imply that the set $\mathbb{R}$ of real numbers should be identified with the subset $\{(x, 0) \mid x \in \mathbb{R}\} \subset \mathbb{C}$. It is also common to use the term **purely imaginary** for any complex number of the form $(0, y)$, where $y \in \mathbb{R}$. In particular, the complex number $i = (0, 1)$ is special, and it is called the **imaginary unit**. (The use of i is standard when denoting this complex number, though j is sometimes used if i means something else. E.g., i is used to denote electric current in Electrical Engineering.)

Note that if we write $1 = (1, 0)$, then we can express $z = (x, y) \in \mathbb{C}$ as

$$z = (x, y) = x(1, 0) + y(0, 1) = x1 + yi = x + yi.$$

It is often significantly easier to perform arithmetic operations on complex numbers when written in this form, as we illustrate in the next section.

2.2   Operations on complex numbers

Even though we have formally defined $\mathbb{C}$ as the set of all ordered pairs of real numbers, we can nonetheless extend the usual arithmetic operations on $\mathbb{R}$ so that they also make sense on $\mathbb{C}$. We discuss such extensions in this section, along with several other important operations on complex numbers.

2.2.1   *Addition and subtraction of complex numbers*

Addition of complex numbers is performed component-wise, meaning that the real and imaginary parts are simply combined.

Definition 2.2.1. Given two complex numbers $(x_1, y_1), (x_2, y_2) \in \mathbb{C}$, we define their **(complex) sum** to be

$$(x_1, y_1) + (x_2, y_2) = (x_1 + x_2, y_1 + y_2).$$

Example 2.2.2. $(3, 2) + (17, -4.5) = (3 + 17, 2 - 4.5) = (20, -2.5)$.

As with the real numbers, subtraction is defined as addition with the so-called **additive inverse**, where the additive inverse of $z = (x, y)$ is defined as $-z = (-x, -y)$.

Example 2.2.3. $(\pi, \sqrt{2}) - (\pi/2, \sqrt{19}) = (\pi, \sqrt{2}) + (-\pi/2, -\sqrt{19})$, where

$$(\pi, \sqrt{2}) + (-\pi/2, -\sqrt{19}) = (\pi - \pi/2, \sqrt{2} - \sqrt{19}) = (\pi/2, \sqrt{2} - \sqrt{19}).$$

The addition of complex numbers shares many of the same properties as the addition of real numbers, including associativity, commutativity, the existence and uniqueness of an additive identity, and the existence and uniqueness of additive inverses. We summarize these properties in the following theorem, which you should prove for your own practice.

Theorem 2.2.4. *Let $z_1, z_2, z_3 \in \mathbb{C}$ be any three complex numbers. Then the following statements are true.*

(1) *(Associativity)* $(z_1 + z_2) + z_3 = z_1 + (z_2 + z_3)$.
(2) *(Commutativity)* $z_1 + z_2 = z_2 + z_1$.
(3) *(Additive Identity) There is a unique complex number, denoted 0, such that, given any complex number $z \in \mathbb{C}$, $0 + z = z$. Moreover, $0 = (0, 0)$.*
(4) *(Additive Inverse) Given any complex number $z \in \mathbb{C}$, there is a unique complex number, denoted $-z$, such that $z + (-z) = 0$. Moreover, if $z = (x, y)$ with $x, y \in \mathbb{R}$, then $-z = (-x, -y)$.*

The proof of this theorem is straightforward and relies solely on the definition of complex addition along with the familiar properties of addition for real numbers.

For example, to check commutativity, let $z_1 = (x_1, y_1)$ and $z_2 = (x_2, y_2)$ be complex numbers with $x_1, x_2, y_1, y_2 \in \mathbb{R}$. Then

$$z_1 + z_2 = (x_1 + x_2, y_1 + y_2) = (x_2 + x_1, y_2 + y_1) = z_2 + z_1.$$

2.2.2 *Multiplication and division of complex numbers*

The definition of multiplication for two complex numbers is at first glance somewhat less straightforward than that of addition.

Definition 2.2.5. Given two complex numbers $(x_1, y_1), (x_2, y_2) \in \mathbb{C}$, we define their **(complex) product** to be

$$(x_1, y_1)(x_2, y_2) = (x_1 x_2 - y_1 y_2, x_1 y_2 + x_2 y_1).$$

According to this definition, $i^2 = -1$. In other words, i is a solution of the polynomial equation $z^2 + 1 = 0$, which does not have solutions in $\mathbb{R}$. Solving such otherwise unsolvable equations was the main motivation behind the introduction of complex numbers. Note that the relation $i^2 = -1$ and the assumption that complex numbers can be multiplied like real numbers is sufficient to arrive at the general rule for multiplication of complex numbers:

$$
\begin{aligned}
(x_1 + y_1 i)(x_2 + y_2 i) &= x_1 x_2 + x_1 y_2 i + x_2 y_1 i + y_1 y_2 i^2 \\
&= x_1 x_2 + x_1 y_2 i + x_2 y_1 i - y_1 y_2 \\
&= x_1 x_2 - y_1 y_2 + (x_1 y_2 + x_2 y_1)i.
\end{aligned}
$$

As with addition, the basic properties of complex multiplication are easy enough to prove using the definition. We summarize these properties in the following theorem, which you should also prove for your own practice.

Theorem 2.2.6. *Let $z_1, z_2, z_3 \in \mathbb{C}$ be any three complex numbers. Then the following statements are true.*

(1) (Associativity) $(z_1 z_2)z_3 = z_1(z_2 z_3)$.
(2) (Commutativity) $z_1 z_2 = z_2 z_1$.
(3) (Multiplicative Identity) There is a unique complex number, denoted 1, such that, given any $z \in \mathbb{C}$, $1z = z$. Moreover, $1 = (1, 0)$.
(4) (Distributivity of Multiplication over Addition) $z_1(z_2 + z_3) = z_1 z_2 + z_1 z_3$.

Just as is the case for real numbers, any non-zero complex number z has a unique multiplicative inverse, which we may denote by either z^{-1} or $1/z$.

Theorem 2.2.6 (continued).

(5) (Multiplicative Inverse) Given $z \in \mathbb{C}$ with $z \neq 0$, there is a unique complex number, denoted z^{-1}, such that $zz^{-1} = 1$. Moreover, if $z = (x, y)$ with $x, y \in \mathbb{R}$, then

$$z^{-1} = \left(\frac{x}{x^2 + y^2}, \frac{-y}{x^2 + y^2} \right).$$

Proof. (*Uniqueness.*) A complex number w is an inverse of z if $zw = 1$ (by the commutativity of complex multiplication this is equivalent to $wz = 1$). We will first prove that if w and v are two complex numbers such that $zw = 1$ and $zv = 1$, then we necessarily have $w = v$. This will then imply that any $z \in \mathbb{C}$ can have at most one inverse. To see this, we start from $zv = 1$. Multiplying both sides by w, we obtain $wzv = w1$. Using the fact that 1 is the multiplicative unit, that the product is commutative, and the assumption that w is an inverse, we get $zwv = v = w$.

(*Existence.*) Now assume $z \in \mathbb{C}$ with $z \neq 0$, and write $z = x + yi$ for $x, y \in \mathbb{R}$. Since $z \neq 0$, at least one of x or y is not zero, and so $x^2 + y^2 > 0$. Therefore, we can define

$$w = \left(\frac{x}{x^2 + y^2}, \frac{-y}{x^2 + y^2} \right),$$

and one can check that $zw = 1$. $\square$

Now, we can define the **division** of a complex number z_1 by a non-zero complex number z_2 as the product of z_1 and z_2^{-1}. Explicitly, for two complex numbers $z_1 = x_1 + iy_1$ and $z_2 = x_2 + iy_2$, $z_2 \neq 0$, we have that their **quotient** is

$$\frac{z_1}{z_2} = \frac{x_1 x_2 + y_1 y_2 + (x_2 y_1 - x_1 y_2)\, i}{x_2^2 + y_2^2}.$$

Example 2.2.7. We illustrate the above definition with the following example:

$$\frac{(1,2)}{(3,4)} = \left(\frac{1 \cdot 3 + 2 \cdot 4}{3^2 + 4^2}, \frac{3 \cdot 2 - 1 \cdot 4}{3^2 + 4^2} \right) = \left(\frac{3+8}{9+16}, \frac{6-4}{9+16} \right) = \left(\frac{11}{25}, \frac{2}{25} \right).$$

2.2.3   *Complex conjugation*

Complex conjugation is an operation on $\mathbb{C}$ that will turn out to be very useful because it allows us to manipulate only the imaginary part of a complex number. In particular, when combined with the notion of modulus (as defined in the next section), it is one of the most fundamental operations on $\mathbb{C}$.

The definition and most basic properties of complex conjugation are as follows. (As in the previous sections, you should provide a proof of the theorem below for your own practice.)

Definition 2.2.8. Given a complex number $z = (x, y) \in \mathbb{C}$ with $x, y \in \mathbb{R}$, we define the **(complex) conjugate** of z to be the complex number

$$\bar{z} = (x, -y).$$

Theorem 2.2.9. *Given two complex numbers $z_1, z_2 \in \mathbb{C}$,*

(1) $\overline{z_1 + z_2} = \overline{z_1} + \overline{z_2}$.
(2) $\overline{z_1 z_2} = \overline{z_1}\, \overline{z_2}$.
(3) $\overline{1/z_1} = 1/\overline{z_1}$, *for all* $z_1 \neq 0$.
(4) $\overline{z_1} = z_1$ *if and only if* $\mathrm{Im}(z_1) = 0$.

(5) $\overline{\overline{z_1}} = z_1$.

(6) the real and imaginary parts of z_1 can be expressed as

$$\mathrm{Re}(z_1) = \frac{1}{2}(z_1 + \overline{z_1}) \quad and \quad \mathrm{Im}(z_1) = \frac{1}{2i}(z_1 - \overline{z_1}).$$

2.2.4 The modulus (a.k.a. norm, length, or magnitude)

In this section, we introduce yet another operation on complex numbers, this time based upon a generalization of the notion of **absolute value** of a real number. To motivate the definition, it is useful to view the set of complex numbers as the two-dimensional Euclidean plane, i.e., to think of $\mathbb{C} = \mathbb{R}^2$ being equal as sets. The **modulus**, or **length**, of $z \in \mathbb{C}$ is then defined as the Euclidean distance between z, as a point in the plane, and the origin $0 = (0,0)$. This is the content of the following definition.

Definition 2.2.10. Given a complex number $z = (x, y) \in \mathbb{C}$ with $x, y \in \mathbb{R}$, the **modulus** of z is defined to be

$$|z| = \sqrt{x^2 + y^2}.$$

In particular, given $x \in \mathbb{R}$, note that

$$|(x, 0)| = \sqrt{x^2 + 0} = |x|$$

under the convention that the square root function takes on its principal positive value.

Example 2.2.11. Using the above definition, we see that the modulus of the complex number $(3, 4)$ is

$$|(3, 4)| = \sqrt{3^2 + 4^2} = \sqrt{9 + 16} = \sqrt{25} = 5.$$

To see this geometrically, construct a figure in the Euclidean plane, such as

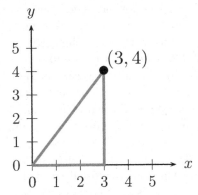

and apply the Pythagorean theorem to the resulting right triangle in order to find the distance from the origin to the point $(3, 4)$.

The following theorem lists the fundamental properties of the modulus, and especially as it relates to complex conjugation. You should provide a proof for your own practice.

Theorem 2.2.12. *Given two complex numbers $z_1, z_2 \in \mathbb{C}$,*

(1) $|z_1 z_2| = |z_1| \cdot |z_2|$.

(2) $\left| \dfrac{z_1}{z_2} \right| = \dfrac{|z_1|}{|z_2|}$, *assuming that $z_2 \neq 0$.*

(3) $|\overline{z_1}| = |z_1|$.

(4) $|\text{Re}(z_1)| \leq |z_1|$ *and* $|\text{Im}(z_1)| \leq |z_1|$.

(5) *(Triangle Inequality)* $|z_1 + z_2| \leq |z_1| + |z_2|$.

(6) *(Another Triangle Inequality)* $|z_1 - z_2| \geq ||z_1| - |z_2||$.

(7) *(Formula for Multiplicative Inverse)* $z_1 \overline{z_1} = |z_1|^2$, *from which*

$$z_1^{-1} = \frac{\overline{z_1}}{|z_1|^2}$$

when we assume that $z_1 \neq 0$.

2.2.5   *Complex numbers as vectors in $\mathbb{R}^2$*

When complex numbers are viewed as points in the Euclidean plane $\mathbb{R}^2$, several of the operations defined in Section 2.2 can be directly visualized as if they were operations on **vectors**.

For the purposes of this Chapter, we think of vectors as directed line segments that start at the origin and end at a specified point in the Euclidean plane. These line segments may also be moved around in space as long as the direction (which we will call the **argument** in Section 2.3.1 below) and the length (a.k.a. the modulus) are preserved. As such, the distinction between points in the plane and vectors is merely a matter of convention as long as we at least implicitly think of each vector as having been translated so that it starts at the origin.

As we saw in Example 2.2.11 above, the modulus of a complex number can be viewed as the length of the hypotenuse of a certain right triangle. The sum and difference of two vectors can also each be represented geometrically as the lengths of specific diagonals within a particular parallelogram that is formed by copying and appropriately translating the two vectors being combined.

Example 2.2.13. We illustrate the sum $(3, 2) + (1, 3) = (4, 5)$ as the main, dashed diagonal of the parallelogram in the left-most figure below. The difference $(3, 2) - (1, 3) = (2, -1)$ can also be viewed as the shorter diagonal of the same parallelogram, though we would properly need to insist that this shorter diagonal be translated so that it starts at the origin. The latter is illustrated in the right-most figure below.

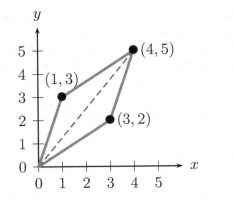
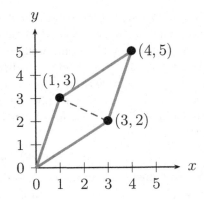

2.3 Polar form and geometric interpretation for $\mathbb{C}$

As mentioned above, $\mathbb{C}$ coincides with the plane $\mathbb{R}^2$ when viewed as a set of ordered pairs of real numbers. Therefore, we can use **polar coordinates** as an alternate way to uniquely identify a complex number. This gives rise to the so-called **polar form** for a complex number, which often turns out to be a convenient representation for complex numbers.

2.3.1 *Polar form for complex numbers*

The following diagram summarizes the relations between cartesian and polar coordinates in $\mathbb{R}^2$:

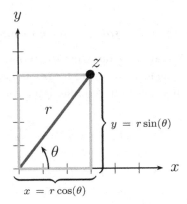

We call the ordered pair (x, y) the **rectangular coordinates** for the complex number z.

We also call the ordered pair (r, θ) the **polar coordinates** for the complex number z. The radius $r = |z|$ is called the **modulus** of z (as defined in Section 2.2.4

above), and the angle $\theta = \mathrm{Arg}(z)$ is called the **argument** of z. Since the argument of a complex number describes an angle that is measured relative to the x-axis, it is important to note that θ is only well-defined up to adding multiples of 2π. As such, we restrict $\theta \in [0, 2\pi)$ and add or subtract multiples of 2π as needed (e.g., when multiplying two complex numbers so that their arguments are added together) in order to keep the argument within this range of values.

It is straightforward to transform polar coordinates into rectangular coordinates using the equations

$$x = r\cos(\theta) \quad \text{and} \quad y = r\sin(\theta). \tag{2.1}$$

In order to transform rectangular coordinates into polar coordinates, we first note that $r = \sqrt{x^2 + y^2}$ is just the complex modulus. Then, θ must be chosen so that it satisfies the bounds $0 \leq \theta < 2\pi$ in addition to the simultaneous equations (2.1) where we are assuming that $z \neq 0$.

Summarizing:

$$z = x + yi = r\cos(\theta) + r\sin(\theta)i = r(\cos(\theta) + \sin(\theta)i).$$

Part of the utility of this expression is that the size $r = |z|$ of z is explicitly part of the very definition since it is easy to check that $|\cos(\theta) + \sin(\theta)i| = 1$ for any choice of $\theta \in \mathbb{R}$.

Closely related is the **exponential form** for complex numbers, which does nothing more than replace the expression $\cos(\theta) + \sin(\theta)i$ with $e^{i\theta}$. The real power of this definition is that this exponential notation turns out to be completely consistent with the usual usage of exponential notation for real numbers.

Example 2.3.1. The complex number i in polar coordinates is expressed as $e^{i\pi/2}$, whereas the number -1 is given by $e^{i\pi}$.

2.3.2   *Geometric multiplication for complex numbers*

As discussed in Section 2.3.1 above, the general exponential form for a complex number z is an expression of the form $re^{i\theta}$ where r is a non-negative real number and $\theta \in [0, 2\pi)$. The utility of this notation is immediately observed when multiplying two complex numbers:

Lemma 2.3.2. *Let $z_1 = r_1 e^{i\theta_1}, z_2 = r_2 e^{i\theta_2} \in \mathbb{C}$ be complex numbers in exponential form. Then*

$$z_1 z_2 = r_1 r_2 e^{i(\theta_1 + \theta_2)}.$$

Proof. By direct computation,

$$\begin{aligned}
z_1 z_2 &= (r_1 e^{i\theta_1})(r_2 e^{i\theta_2}) = r_1 r_2 e^{i\theta_1} e^{i\theta_2} \\
&= r_1 r_2 (\cos\theta_1 + i\sin\theta_1)(\cos\theta_2 + i\sin\theta_2) \\
&= r_1 r_2 \big[(\cos\theta_1 \cos\theta_2 - \sin\theta_1 \sin\theta_2) + i(\sin\theta_1 \cos\theta_2 + \cos\theta_1 \sin\theta_2)\big] \\
&= r_1 r_2 \big[\cos(\theta_1 + \theta_2) + i\sin(\theta_1 + \theta_2)\big] = r_1 r_2 e^{i(\theta_1 + \theta_2)},
\end{aligned}$$

where we have used the usual formulas for the sine and cosine of the sum of two angles. □

In particular, Lemma 2.3.2 shows that the modulus $|z_1 z_2|$ of the product is the product of the moduli r_1 and r_2 and that the argument $\mathrm{Arg}(z_1 z_2)$ of the product is the sum of the arguments $\theta_1 + \theta_2$.

2.3.3 *Exponentiation and root extraction*

Another important use for the polar form of a complex number is in exponentiation. The simplest possible situation here involves the use of a positive integer as a power, in which case exponentiation is nothing more than repeated multiplication. Given the observations in Section 2.3.2 above and using some trigonometric identities, one quickly obtains the following fundamental result.

Theorem 2.3.3 (de Moivre's Formula). *Let $z = r(\cos(\theta) + \sin(\theta)i)$ be a complex number in polar form and $n \in \mathbb{Z}_+$ be a positive integer. Then*

(1) the exponentiation $z^n = r^n(\cos(n\theta) + \sin(n\theta)i)$ and
(2) the nth roots of z are given by the n complex numbers

$$z_k = r^{1/n}\left[\cos\left(\frac{\theta}{n} + \frac{2\pi k}{n}\right) + \sin\left(\frac{\theta}{n} + \frac{2\pi k}{n}\right)i\right] = r^{1/n}e^{\frac{i}{n}(\theta + 2\pi k)},$$

where $k = 0, 1, 2, \ldots, n-1$.

Note, in particular, that we are not only always guaranteed the existence of an nth root for any complex number, but that we are also always guaranteed to have exactly n of them. This level of completeness in root extraction is in stark contrast with roots of real numbers (within the real numbers) which may or may not exist and may be unique or not when they exist.

An important special case of de Moivre's Formula yields n n^{th} **roots of unity**. By **unity**, we just mean the complex number $1 = 1 + 0i$, and by the nth **roots of unity**, we mean the n numbers

$$z_k = 1^{1/n}\left[\cos\left(\frac{0}{n} + \frac{2\pi k}{n}\right) + \sin\left(\frac{0}{n} + \frac{2\pi k}{n}\right)i\right]$$

$$= \cos\left(\frac{2\pi k}{n}\right) + \sin\left(\frac{2\pi k}{n}\right)i$$

$$= e^{2\pi i(k/n)},$$

where $k = 0, 1, 2, \ldots, n-1$. The fact that these numbers are precisely the complex numbers solving the equation $z^n = 1$, has many interesting applications.

Example 2.3.4. To find all solutions of the equation $z^3 + 8 = 0$ for $z \in \mathbb{C}$, we may write $z = re^{i\theta}$ in polar form with $r > 0$ and $\theta \in [0, 2\pi)$. Then the equation $z^3 + 8 = 0$ becomes $z^3 = r^3 e^{i3\theta} = -8 = 8e^{i\pi}$ so that $r = 2$ and $3\theta = \pi + 2\pi k$

for $k = 0, 1, 2$. This means that there are three distinct solutions when $\theta \in [0, 2\pi)$, namely $\theta = \frac{\pi}{3}$, $\theta = \pi$, and $\theta = \frac{5\pi}{3}$.

2.3.4 *Some complex elementary functions*

We conclude this chapter by defining three of the basic elementary functions that take complex arguments. In this context, "elementary function" is used as a technical term and essentially means something like "one of the most common forms of functions encountered when beginning to learn Calculus." The most basic elementary functions include the familiar polynomial and algebraic functions, such as the nth root function, in addition to the somewhat more sophisticated exponential function, the trigonometric functions, and the logarithmic function. For the purposes of this chapter, we will now define the complex exponential function and two complex trigonometric functions. Definitions for the remaining basic elementary functions can be found in any book on Complex Analysis.

The basic groundwork for defining the **complex exponential function** was already put into place in Sections 2.3.1 and 2.3.2 above. In particular, we have already defined the expression $e^{i\theta}$ to mean the sum $\cos(\theta) + \sin(\theta)i$ for any real number θ. Historically, this equivalence is a special case of the more general **Euler's formula**

$$e^{x+yi} = e^x(\cos(y) + \sin(y)i),$$

which we here take as our definition of the complex exponential function applied to any complex number $x + yi$ for $x, y \in \mathbb{R}$.

Given this exponential function, one can then define the **complex sine function** and the **complex cosine function** as

$$\sin(z) = \frac{e^{iz} - e^{-iz}}{2i} \quad \text{and} \quad \cos(z) = \frac{e^{iz} + e^{-iz}}{2}.$$

Remarkably, these functions retain many of their familiar properties, which should be taken as a sign that the definitions — however abstract — have been well thought-out. We summarize a few of these properties as follows.

Theorem 2.3.5. *Given $z_1, z_2 \in \mathbb{C}$,*

(1) $e^{z_1 + z_2} = e^{z_1} e^{z_2}$ and $e^z \neq 0$ for any choice of $z \in \mathbb{C}$.
(2) $\sin^2(z_1) + \cos^2(z_1) = 1$.
(3) $\sin(z_1 + z_2) = \sin(z_1) \cdot \cos(z_2) + \cos(z_1) \cdot \sin(z_2)$.
(4) $\cos(z_1 + z_2) = \cos(z_1) \cdot \cos(z_2) - \sin(z_1) \cdot \sin(z_2)$.

Exercises for Chapter 2

Calculational Exercises

(1) Express the following complex numbers in the form $x + yi$ for $x, y \in \mathbb{R}$:

    (a) $(2 + 3i) + (4 + i)$

    (b) $(2 + 3i)^2(4 + i)$

    (c) $\dfrac{2 + 3i}{4 + i}$

    (d) $\dfrac{1}{i} + \dfrac{3}{1 + i}$

    (e) $(-i)^{-1}$

    (f) $(-1 + i\sqrt{3})^3$

(2) Compute the real and imaginary parts of the following expressions, where z is the complex number $x + yi$ and $x, y \in \mathbb{R}$:

    (a) $\dfrac{1}{z^2}$

    (b) $\dfrac{1}{3z + 2}$

    (c) $\dfrac{z + 1}{2z - 5}$

    (d) z^3

(3) Find $r > 0$ and $\theta \in [0, 2\pi)$ such that $(1 - i)/\sqrt{2} = re^{i\theta}$.

(4) Solve the following equations for z a complex number:

    (a) $z^5 - 2 = 0$

    (b) $z^4 + i = 0$

    (c) $z^6 + 8 = 0$

    (d) $z^3 - 4i = 0$

(5) Calculate the

    (a) complex conjugate of the fraction $(3 + 8i)^4/(1 + i)^{10}$.

    (b) complex conjugate of the fraction $(8 - 2i)^{10}/(4 + 6i)^5$.

    (c) complex modulus of the fraction $i(2 + 3i)(5 - 2i)/(-2 - i)$.

    (d) complex modulus of the fraction $(2 - 3i)^2/(8 + 6i)^2$.

(6) Compute the real and imaginary parts:

    (a) e^{2+i}

    (b) $\sin(1 + i)$

    (c) e^{3-i}

    (d) $\cos(2 + 3i)$

(7) Compute the real and imaginary parts of e^{e^z} for $z \in \mathbb{C}$.

Proof-Writing Exercises

(1) Let $a \in \mathbb{R}$ and $z, w \in \mathbb{C}$. Prove that

    (a) $\mathrm{Re}(az) = a\mathrm{Re}(z)$ and $\mathrm{Im}(az) = a\mathrm{Im}(z)$.

    (b) $\mathrm{Re}(z + w) = \mathrm{Re}(z) + \mathrm{Re}(w)$ and $\mathrm{Im}(z + w) = \mathrm{Im}(z) + \mathrm{Im}(w)$.

(2) Let $z \in \mathbb{C}$. Prove that $\mathrm{Im}(z) = 0$ if and only if $\mathrm{Re}(z) = z$.

(3) Let $z, w \in \mathbb{C}$. Prove the *parallelogram law* $|z - w|^2 + |z + w|^2 = 2(|z|^2 + |w|^2)$.

(4) Let $z, w \in \mathbb{C}$ with $\overline{z}w \neq 1$ such that either $|z| = 1$ or $|w| = 1$. Prove that
$$\left| \frac{z - w}{1 - \overline{z}w} \right| = 1.$$

(5) For an angle $\theta \in [0, 2\pi)$, find the linear map $f_\theta : \mathbb{R}^2 \to \mathbb{R}^2$, which describes the rotation by the angle θ in the counterclockwise direction.

Hint: For a given angle θ, find $a, b, c, d \in \mathbb{R}$ such that $f_\theta(x_1, x_2) = (ax_1 + bx_2, cx_1 + dx_2)$.

Chapter 3

The Fundamental Theorem of Algebra and Factoring Polynomials

The similarities and differences between $\mathbb{R}$ and $\mathbb{C}$ are elegant and intriguing, but why are complex numbers important? One possible answer to this question is the **Fundamental Theorem of Algebra**. It states that every polynomial equation in one variable with complex coefficients has at least one complex solution. In other words, polynomial equations formed over $\mathbb{C}$ can always be solved over $\mathbb{C}$. This amazing result has several equivalent formulations in addition to a myriad of different proofs, one of the first of which was given by the eminent mathematician Carl Friedrich Gauss (1777-1855) in his doctoral thesis.

3.1 The Fundamental Theorem of Algebra

The aim of this section is to provide a proof of the Fundamental Theorem of Algebra using concepts that should be familiar from the study of Calculus, and so we begin by providing an explicit formulation.

Theorem 3.1.1 (Fundamental Theorem of Algebra). *Given any positive integer $n \in \mathbb{Z}_+$ and any choice of complex numbers $a_0, a_1, \ldots, a_n \in \mathbb{C}$ with $a_n \neq 0$, the polynomial equation*

$$a_n z^n + \cdots + a_1 z + a_0 = 0 \tag{3.1}$$

has at least one solution $z \in \mathbb{C}$.

This is a remarkable statement. No analogous result holds for guaranteeing that a real solution exists to Equation (3.1) if we restrict the coefficients $a_0, a_1, \ldots, a_n$ to be real numbers. E.g., there does not exist a real number x satisfying an equation as simple as $\pi x^2 + e = 0$. Similarly, the consideration of polynomial equations having integer (resp. rational) coefficients quickly forces us to consider solutions that cannot possibly be integers (resp. rational numbers). Thus, the complex numbers are special in this respect.

The statement of the Fundamental Theorem of Algebra can also be read as follows: Any non-constant complex polynomial function defined on the complex plane $\mathbb{C}$ (when thought of as $\mathbb{R}^2$) has at least one root, i.e., vanishes in at least one place. It is in this form that we will provide a proof for Theorem 3.1.1.

Given how long the Fundamental Theorem of Algebra has been around, you should not be surprised that there are many proofs of it. There have even been entire books devoted solely to exploring the mathematics behind various distinct proofs. Different proofs arise from attempting to understand the statement of the theorem from the viewpoint of different branches of mathematics. This quickly leads to many non-trivial interactions with such fields of mathematics as Real and Complex Analysis, Topology, and (Modern) Abstract Algebra. The diversity of proof techniques available is yet another indication of how fundamental and deep the Fundamental Theorem of Algebra really is.

To prove the Fundamental Theorem of Algebra using Differential Calculus, we will need the **Extreme Value Theorem** for real-valued functions of two real variables, which we state without proof. In particular, we formulate this theorem in the restricted case of functions defined on the **closed disk** D of radius $R > 0$ and centered at the origin, i.e.,

$$D = \{(x_1, x_2) \in \mathbb{R}^2 \mid x_1^2 + x_2^2 \leq R^2\}.$$

Theorem 3.1.2 (Extreme Value Theorem). *Let $f : D \to \mathbb{R}$ be a continuous function on the closed disk $D \subset \mathbb{R}^2$. Then f is bounded and attains its minimum and maximum values on D. In other words, there exist points $x_m, x_M \in D$ such that*

$$f(x_m) \leq f(x) \leq f(x_M)$$

for every possible choice of point $x \in D$.

If we define a polynomial function $f : \mathbb{C} \to \mathbb{C}$ by setting $f(z) = a_n z^n + \cdots + a_1 z + a_0$ as in Equation (3.1), then note that we can regard $(x, y) \mapsto |f(x + iy)|$ as a function $\mathbb{R}^2 \to \mathbb{R}$. By a mild abuse of notation, we denote this function by $|f(\cdot)|$ or $|f|$. As it is a composition of continuous functions (polynomials and the square root), we see that $|f|$ is also continuous.

Lemma 3.1.3. *Let $f : \mathbb{C} \to \mathbb{C}$ be any polynomial function. Then there exists a point $z_0 \in \mathbb{C}$ where the function $|f|$ attains its minimum value in $\mathbb{R}$.*

Proof. If f is a constant polynomial function, then the statement of the Lemma is trivially true since $|f|$ attains its minimum value at every point in $\mathbb{C}$. So choose, e.g., $z_0 = 0$.

If f is not constant, then the degree of the polynomial defining f is at least one. In this case, we can denote f explicitly as in Equation (3.1). That is, we set

$$f(z) = a_n z^n + \cdots + a_1 z + a_0$$

with $a_n \neq 0$. Now, assume $z \neq 0$, and set $A = \max\{|a_0|, \dots, |a_{n-1}|\}$. We can obtain a lower bound for $|f(z)|$ as follows:

$$|f(z)| = |a_n| \, |z|^n \left| 1 + \frac{a_{n-1}}{a_n} \frac{1}{z} + \dots + \frac{a_0}{a_n} \frac{1}{z^n} \right|$$

$$\geq |a_n| \, |z|^n \left(1 - \frac{A}{|a_n|} \sum_{k=1}^{\infty} \frac{1}{|z|^k} \right) = |a_n| \, |z|^n \left(1 - \frac{A}{|a_n|} \frac{1}{|z| - 1} \right).$$

For all $z \in \mathbb{C}$ such that $|z| \geq 2$, we can further simplify this expression and obtain

$$|f(z)| \geq |a_n| \, |z|^n \left(1 - \frac{2A}{|a_n| |z|} \right).$$

It follows from this inequality that there is an $R > 0$ such that $|f(z)| > |f(0)|$, for all $z \in \mathbb{C}$ satisfying $|z| > R$. Let $D \subset \mathbb{R}^2$ be the disk of radius R centered at 0, and define a function $g : D \to \mathbb{R}$, by

$$g(x, y) = |f(x + iy)|.$$

Since g is continuous, we can apply Theorem 3.1.2 in order to obtain a point $(x_0, y_0) \in D$ such that g attains its minimum at (x_0, y_0). By the choice of R we have that for $z \in \mathbb{C} \setminus D$, $|f(z)| > |g(0,0)| \geq |g(x_0, y_0)|$. Therefore, $|f|$ attains its minimum at $z = x_0 + iy_0$. $\qquad \square$

We now prove the Fundamental Theorem of Algebra.

Proof of Theorem 3.1.1. For our argument, we rely on the fact that the function $|f|$ attains its minimum value by Lemma 3.1.3. Let $z_0 \in \mathbb{C}$ be a point where the minimum is attained. We will show that if $f(z_0) \neq 0$, then z_0 is *not* a minimum, thus proving by contraposition that the minimum value of $|f(z)|$ is zero. Therefore, $f(z_0) = 0$.

If $f(z_0) \neq 0$, then we can define a new function $g : \mathbb{C} \to \mathbb{C}$ by setting

$$g(z) = \frac{f(z + z_0)}{f(z_0)}, \quad \text{for all } z \in \mathbb{C}.$$

Note that g is a polynomial of degree n, and that the minimum of $|f|$ is attained at z_0 if and only if the minimum of $|g|$ is attained at $z = 0$. Moreover, it is clear that $g(0) = 1$.

More explicitly, g is given by a polynomial of the form

$$g(z) = b_n z^n + \dots + b_k z^k + 1,$$

with $n \geq 1$ and $b_k \neq 0$, for some $1 \leq k \leq n$. Let $b_k = |b_k| e^{i\theta}$, and consider z of the form

$$z = r |b_k|^{-1/k} e^{i(\pi - \theta)/k}, \tag{3.2}$$

with $r > 0$. For z of this form we have

$$g(z) = 1 - r^k + r^{k+1} h(r),$$

where h is a polynomial. Then, for $r < 1$, we have by the triangle inequality that

$$|g(z)| \leq 1 - r^k + r^{k+1}|h(r)|.$$

For $r > 0$ sufficiently small we have $r|h(r)| < 1$, by the continuity of the function $rh(r)$ and the fact that it vanishes in $r = 0$. Hence

$$|g(z)| \leq 1 - r^k(1 - r|h(r)|) < 1,$$

for some z having the form in Equation (3.2) with $r \in (0, r_0)$ and $r_0 > 0$ sufficiently small. But then the minimum of the function $|g| : \mathbb{C} \to \mathbb{R}$ cannot possibly be equal to 1. $\qquad\square$

3.2   Factoring polynomials

In this section, we present several fundamental facts about polynomials, including an equivalent form of the Fundamental Theorem of Algebra. While these facts should be familiar, they nonetheless require careful formulation and proof. Before stating these results, though, we first present a review of the main concepts needed in order to more carefully work with polynomials.

Let $n \in \mathbb{Z}_+ \cup \{0\}$ be a non-negative integer, and let $a_0, a_1, \ldots, a_n \in \mathbb{C}$ be complex numbers. Then we call the expression

$$p(z) = a_n z^n + \cdots + a_1 z + a_0$$

a **polynomial** in the variable z with **coefficients** $a_0, a_1, \ldots, a_n$. If $a_n \neq 0$, then we say that $p(z)$ has **degree** n (denoted $\deg(p(z)) = n$), and we call a_n the **leading term** of $p(z)$. Moreover, if $a_n = 1$, then we call $p(z)$ a **monic polynomial**. If, however, $n = a_0 = 0$, then we call $p(z) = 0$ the **zero polynomial** and set $\deg(0) = -\infty$.

Finally, by a **root** (a.k.a. **zero**) of a polynomial $p(z)$, we mean a complex number z_0 such that $p(z_0) = 0$. Note, in particular, that every complex number is a root of the zero polynomial.

Convention dictates that

- a degree zero polynomial be called a **constant polynomial**,
- a degree one polynomial be called a **linear polynomial**,
- a degree two polynomial be called a **quadratic polynomial**,
- a degree three polynomial be called a **cubic polynomial**,
- a degree four polynomial be called a **quadric polynomial**,
- a degree five polynomial be called a **quintic polynomial**,
- and so on.

Addition and multiplication of polynomials is a direct generalization of the addition and multiplication of complex numbers, and degree interacts with these operations as follows:

Lemma 3.2.1. *Let $p(z)$ and $q(z)$ be non-zero polynomials. Then*

(1) $\deg\left(p(z) \pm q(z)\right) \leq \max\{\deg(p(z)), \deg(q(z))\}$
(2) $\deg\left(p(z)q(z)\right) = \deg(p(z)) + \deg(q(z))$.

Theorem 3.2.2. *Given a positive integer $n \in \mathbb{Z}_+$ and any choice of $a_0, a_1, \ldots, a_n \in \mathbb{C}$ with $a_n \neq 0$, define the function $f : \mathbb{C} \to \mathbb{C}$ by setting*

$$f(z) = a_n z^n + \cdots + a_1 z + a_0, \forall z \in \mathbb{C}.$$

In other words, f is a polynomial function of degree n. Then

(1) *given any complex number $w \in \mathbb{C}$, we have that $f(w) = 0$ if and only if there exists a polynomial function $g : \mathbb{C} \to \mathbb{C}$ of degree $n - 1$ such that*

$$f(z) = (z - w)g(z), \forall z \in \mathbb{C}.$$

(2) *there are at most n distinct complex numbers w for which $f(w) = 0$. In other words, f has at most n distinct roots.*

(3) *(Fundamental Theorem of Algebra, restated) there exist exactly $n + 1$ complex numbers $w_0, w_1, \ldots, w_n \in \mathbb{C}$ (not necessarily distinct) such that*

$$f(z) = w_0(z - w_1)(z - w_2) \cdots (z - w_n), \forall z \in \mathbb{C}.$$

In other words, every polynomial function with coefficients over $\mathbb{C}$ can be factored into linear factors over $\mathbb{C}$.

Proof.

(1) Let $w \in \mathbb{C}$ be a complex number.

("$\Longrightarrow$") Suppose that $f(w) = 0$. Then, in particular, we have that

$$a_n w^n + \cdots + a_1 w + a_0 = 0.$$

Since this equation is equal to zero, it follows that, given any $z \in \mathbb{C}$,

$$
\begin{aligned}
f(z) &= a_n z^n + \cdots + a_1 z + a_0 - (a_n w^n + \cdots + a_1 w + a_0) \\
&= a_n(z^n - w^n) + a_{n-1}(z^{n-1} - w^{n-1}) + \cdots + a_1(z - w) \\
&= a_n(z - w) \sum_{k=0}^{n-1} z^k w^{n-1-k} + a_{n-1}(z - w) \sum_{k=0}^{n-2} z^k w^{n-2-k} + \cdots + a_1(z - w) \\
&= (z - w) \sum_{m=1}^{n} \left(a_m \sum_{k=0}^{m-1} z^k w^{m-1-k} \right).
\end{aligned}
$$

Thus, upon setting

$$g(z) = \sum_{m=1}^{n} \left(a_m \sum_{k=0}^{m-1} z^k w^{m-1-k} \right), \forall z \in \mathbb{C},$$

we have constructed a degree $n - 1$ polynomial function g such that

$$f(z) = (z - w)g(z), \forall z \in \mathbb{C}.$$

("$\Longleftarrow$") Suppose that there exists a polynomial function $g : \mathbb{C} \to \mathbb{C}$ of degree $n - 1$ such that

$$f(z) = (z - w)g(z), \forall z \in \mathbb{C}.$$

Then it follows that $f(w) = (w - w)g(w) = 0$, as desired.

(2) We use induction on the degree n of f.

If $n = 1$, then $f(z) = a_1 z + a_0$ is a linear function, and the equation $a_1 z + a_0 = 0$ has the unique solution $z = -a_0/a_1$. Thus, the result holds for $n = 1$.

Now, suppose that the result holds for $n - 1$. In other words, assume that every polynomial function of degree $n - 1$ has at most $n - 1$ roots. Using the Fundamental Theorem of Algebra (Theorem 3.1.1), we know that there exists a complex number $w \in \mathbb{C}$ such that $f(w) = 0$. Moreover, from Part 1 above, we know that there exists a polynomial function g of degree $n - 1$ such that

$$f(z) = (z - w)g(z), \forall z \in \mathbb{C}.$$

It then follows by the induction hypothesis that g has at most $n - 1$ distinct roots, and so f must have at most n distinct roots.

(3) This part follows from an induction argument on n that is virtually identical to that of Part 2, and so the proof is left as an exercise to the reader.

$\square$

Exercises for Chapter 3

Calculational Exercises

(1) Let $n \in \mathbb{Z}_+$ be a positive integer, let $w_0, w_1, \ldots, w_n \in \mathbb{C}$ be distinct complex numbers, and let $z_0, z_1, \ldots, z_n \in \mathbb{C}$ be any complex numbers. Then one can prove that there is a unique polynomial $p(z)$ of degree at most n such that, for each $k \in \{0, 1, \ldots, n\}$, $p(w_k) = z_k$.

  (a) Find the unique polynomial of degree at most 2 that satisfies $p(0) = 0$, $p(1) = 1$, and $p(2) = 2$.

  (b) Can your result in Part (a) be easily generalized to find the unique polynomial of degree at most n satisfying $p(0) = 0$, $p(1) = 1$, $\ldots$, $p(n) = n$?

(2) Given any complex number $\alpha \in \mathbb{C}$, show that the coefficients of the polynomial

$$(z - \alpha)(z - \overline{\alpha})$$

are real numbers.

Proof-Writing Exercises

(1) Let $m, n \in \mathbb{Z}_+$ be positive integers with $m \leq n$. Prove that there is a degree n polynomial $p(z)$ with complex coefficients such that $p(z)$ has exactly m distinct roots.

(2) Given a polynomial $p(z) = a_n z^n + \cdots + a_1 z + a_0$ with complex coefficients, define the **conjugate** of $p(z)$ to be the new polynomial

$$\overline{p}(z) = \overline{a_n} z^n + \cdots + \overline{a_1} z + \overline{a_0}.$$

(a) Prove that $\overline{p(z)} = \overline{p}(\overline{z})$.

(b) Prove that $p(z)$ has real coefficients if and only if $\overline{p}(z) = p(z)$.

(c) Given polynomials $p(z)$, $q(z)$, and $r(z)$ such that $p(z) = q(z)r(z)$, prove that $\overline{p}(z) = \overline{q}(z)\overline{r}(z)$.

(3) Let $p(z)$ be a polynomial with real coefficients, and let $\alpha \in \mathbb{C}$ be a complex number.

Prove that $p(\alpha) = 0$ if and only $p(\overline{\alpha}) = 0$.

Chapter 4

Vector Spaces

With the background developed in the previous chapters, we are ready to begin the study of Linear Algebra by introducing vector spaces. Vector spaces are essential for the formulation and solution of linear algebra problems and they will appear on virtually every page of this book from now on.

4.1 Definition of vector spaces

As we have seen in Chapter 1, a vector space is a set V with two operations defined upon it: addition of vectors and multiplication by scalars. These operations must satisfy certain properties, which we are about to discuss in more detail. The scalars are taken from a field $\mathbb{F}$, where for the remainder of this book $\mathbb{F}$ stands either for the real numbers $\mathbb{R}$ or for the complex numbers $\mathbb{C}$. The sets $\mathbb{R}$ and $\mathbb{C}$ are examples of fields. The abstract definition of a field along with further examples can be found in Appendix C.

Vector addition can be thought of as a function $+ : V \times V \to V$ that maps two vectors $u, v \in V$ to their sum $u + v \in V$. **Scalar multiplication** can similarly be described as a function $\mathbb{F} \times V \to V$ that maps a scalar $a \in \mathbb{F}$ and a vector $v \in V$ to a new vector $av \in V$. (More information on these kinds of functions, also known as binary operations, can be found in Appendix C.) It is when we place the right conditions on these operations, also called **axioms**, that we turn V into a vector space.

Definition 4.1.1. A **vector space** over $\mathbb{F}$ is a set V together with the operations of addition $V \times V \to V$ and scalar multiplication $\mathbb{F} \times V \to V$ satisfying each of the following properties.

(1) **Commutativity:** $u + v = v + u$ for all $u, v \in V$;
(2) **Associativity:** $(u + v) + w = u + (v + w)$ and $(ab)v = a(bv)$ for all $u, v, w \in V$ and $a, b \in \mathbb{F}$;
(3) **Additive identity:** There exists an element $0 \in V$ such that $0 + v = v$ for all $v \in V$;

(4) **Additive inverse:** For every $v \in V$, there exists an element $w \in V$ such that $v + w = 0$;
(5) **Multiplicative identity:** $1v = v$ for all $v \in V$;
(6) **Distributivity:** $a(u + v) = au + av$ and $(a + b)u = au + bu$ for all $u, v \in V$ and $a, b \in \mathbb{F}$.

A vector space over $\mathbb{R}$ is usually called a **real vector space**, and a vector space over $\mathbb{C}$ is similarly called a **complex vector space**. The elements $v \in V$ of a vector space are called **vectors**.

Even though Definition 4.1.1 may appear to be an extremely abstract definition, vector spaces are fundamental objects in mathematics because there are countless examples of them. You should expect to see many examples of vector spaces throughout your mathematical life.

Example 4.1.2. Consider the set $\mathbb{F}^n$ of all n-tuples with elements in $\mathbb{F}$. This is a vector space with addition and scalar multiplication defined componentwise. That is, for $u = (u_1, u_2, \ldots, u_n), v = (v_1, v_2, \ldots, v_n) \in \mathbb{F}^n$ and $a \in \mathbb{F}$, we define

$$u + v = (u_1 + v_1, u_2 + v_2, \ldots, u_n + v_n),$$
$$au = (au_1, au_2, \ldots, au_n).$$

It is easy to check that each property of Definition 4.1.1 is satisfied. In particular, the additive identity $0 = (0, 0, \ldots, 0)$, and the additive inverse of u is $-u = (-u_1, -u_2, \ldots, -u_n)$.

An important case of Example 4.1.2 is $\mathbb{R}^n$, especially when $n = 2$ or $n = 3$. We have already seen in Chapter 1 that there is a geometric interpretation for elements of $\mathbb{R}^2$ and $\mathbb{R}^3$ as points in the Euclidean plane and Euclidean space, respectively.

Example 4.1.3. Let $\mathbb{F}^\infty$ be the set of all sequences over $\mathbb{F}$, i.e.,

$$\mathbb{F}^\infty = \{(u_1, u_2, \ldots) \mid u_j \in \mathbb{F} \text{ for } j = 1, 2, \ldots\}.$$

Addition and scalar multiplication are defined as expected, namely,

$$(u_1, u_2, \ldots) + (v_1, v_2, \ldots) = (u_1 + v_1, u_2 + v_2, \ldots),$$
$$a(u_1, u_2, \ldots) = (au_1, au_2, \ldots).$$

You should verify that $\mathbb{F}^\infty$ becomes a vector space under these operations.

Example 4.1.4. Verify that $V = \{0\}$ is a vector space! (Here, 0 denotes the zero vector in any vector space.)

Example 4.1.5. Let $\mathbb{F}[z]$ be the set of all polynomial functions $p : \mathbb{F} \to \mathbb{F}$ with coefficients in $\mathbb{F}$. As discussed in Chapter 3, $p(z)$ is a polynomial if there exist $a_0, a_1, \ldots, a_n \in \mathbb{F}$ such that

$$p(z) = a_n z^n + a_{n-1} z^{n-1} + \cdots + a_1 z + a_0. \tag{4.1}$$

Addition and scalar multiplication on $\mathbb{F}[z]$ are defined pointwise as

$$(p + q)(z) = p(z) + q(z),$$
$$(ap)(z) = ap(z),$$

where $p, q \in \mathbb{F}[z]$ and $a \in \mathbb{F}$. For example, if $p(z) = 5z + 1$ and $q(z) = 2z^2 + z + 1$, then $(p + q)(z) = 2z^2 + 6z + 2$ and $(2p)(z) = 10z + 2$.

It can be easily verified that, under these operations, $\mathbb{F}[z]$ forms a vector space over $\mathbb{F}$. The additive identity in this case is the zero polynomial, for which all coefficients are equal to zero, and the additive inverse of $p(z)$ in Equation (4.1) is $-p(z) = -a_n z^n - a_{n-1} z^{n-1} - \cdots - a_1 z - a_0$.

Example 4.1.6. Extending Example 4.1.5, let $D \subset \mathbb{R}$ be a subset of $\mathbb{R}$, and let $\mathcal{C}(D)$ denote the set of all continuous functions with domain D and codomain $\mathbb{R}$. Then, under the same operations of pointwise addition and scalar multiplication, one can show that $\mathcal{C}(D)$ also forms a vector space.

4.2 Elementary properties of vector spaces

We are going to prove several important, yet simple, properties of vector spaces. From now on, V will denote a vector space over $\mathbb{F}$.

Proposition 4.2.1. *In every vector space the additive identity is unique.*

Proof. Suppose there are two additive identities 0 and $0'$. Then

$$0' = 0 + 0' = 0,$$

where the first equality holds since 0 is an identity and the second equality holds since $0'$ is an identity. Hence $0 = 0'$, proving that the additive identity is unique. $\square$

Proposition 4.2.2. *Every $v \in V$ has a unique additive inverse.*

Proof. Suppose w and w' are additive inverses of v so that $v + w = 0$ and $v + w' = 0$. Then

$$w = w + 0 = w + (v + w') = (w + v) + w' = 0 + w' = w'.$$

Hence $w = w'$, as desired. $\square$

Since the additive inverse of v is unique, as we have just shown, it will from now on be denoted by $-v$. We also define $w - v$ to mean $w + (-v)$. We will, in fact, show in Proposition 4.2.5 below that $-v = -1v$.

Proposition 4.2.3. $0v = 0$ *for all* $v \in V$.

Note that the 0 on the left-hand side in Proposition 4.2.3 is a scalar, whereas the 0 on the right-hand side is a vector.

Proof. For $v \in V$, we have by distributivity that

$$0v = (0 + 0)v = 0v + 0v.$$

Adding the additive inverse of $0v$ to both sides, we obtain

$$0 = 0v - 0v = (0v + 0v) - 0v = 0v.$$

□

Proposition 4.2.4. $a0 = 0$ *for every* $a \in \mathbb{F}$.

Proof. As in the proof of Proposition 4.2.3, if $a \in \mathbb{F}$, then

$$a0 = a(0 + 0) = a0 + a0.$$

Adding the additive inverse of $a0$ to both sides, we obtain $0 = a0$, as desired.   □

Proposition 4.2.5. $(-1)v = -v$ *for every* $v \in V$.

Proof. For $v \in V$, we have

$$v + (-1)v = 1v + (-1)v = (1 + (-1))v = 0v = 0,$$

which shows that $(-1)v$ is the additive inverse $-v$ of v.   □

4.3   Subspaces

As mentioned in the last section, there are countless examples of vector spaces. One particularly important source of new vector spaces comes from looking at subsets of a set that is already known to be a vector space.

Definition 4.3.1. Let V be a vector space over $\mathbb{F}$, and let $U \subset V$ be a subset of V. Then we call U a **subspace** of V if U is a vector space over $\mathbb{F}$ under the same operations that make V into a vector space over $\mathbb{F}$.

To check that a subset $U \subset V$ is a subspace, it suffices to check only a few of the conditions of a vector space.

Lemma 4.3.2. *Let* $U \subset V$ *be a subset of a vector space* V *over* $\mathbb{F}$. *Then* U *is a subspace of* V *if and only if the following three conditions hold.*

(1) **additive identity:** $0 \in U$*;*
(2) **closure under addition:** $u, v \in U$ *implies* $u + v \in U$*;*
(3) **closure under scalar multiplication:** $a \in \mathbb{F}$, $u \in U$ *implies that* $au \in U$*.*

Proof. Condition 1 implies that the additive identity exists. Condition 2 implies that vector addition is well-defined and, Condition 3 ensures that scalar multiplication is well-defined. All other conditions for a vector space are inherited from V since addition and scalar multiplication for elements in U are the same when viewed as elements in either U or V.   □

Remark 4.3.3. Note that if we require $U \subset V$ to be a *nonempty* subset of V, then condition 1 of Lemma 4.3.2 already follows from condition 3 since $0u = 0$ for $u \in U$.

Example 4.3.4. In every vector space V, the subsets $\{0\}$ and V are easily verified to be subspaces. We call these the **trivial subspaces** of V.

Example 4.3.5. $\{(x_1, 0) \mid x_1 \in \mathbb{R}\}$ is a subspace of $\mathbb{R}^2$.

Example 4.3.6. $U = \{(x_1, x_2, x_3) \in \mathbb{F}^3 \mid x_1 + 2x_2 = 0\}$ is a subspace of $\mathbb{F}^3$. To see this, we need to check the three conditions of Lemma 4.3.2.

The zero vector $(0, 0, 0) \in \mathbb{F}^3$ is in U since it satisfies the condition $x_1 + 2x_2 = 0$. To show that U is closed under addition, take two vectors $v = (v_1, v_2, v_3)$ and $u = (u_1, u_2, u_3)$. Then, by the definition of U, we have $v_1 + 2v_2 = 0$ and $u_1 + 2u_2 = 0$. Adding these two equations, it is not hard to see that the vector $v + u = (v_1 + u_1, v_2 + u_2, v_3 + u_3)$ satisfies $(v_1 + u_1) + 2(v_2 + u_2) = 0$. Hence $v + u \in U$. Similarly, to show closure under scalar multiplication, take $u = (u_1, u_2, u_3) \in U$ and $a \in \mathbb{F}$. Then $au = (au_1, au_2, au_3)$ satisfies the equation $au_1 + 2au_2 = a(u_1 + 2u_2) = 0$, and so $au \in U$.

Example 4.3.7. $U = \{p \in \mathbb{F}[z] \mid p(3) = 0\}$ is a subspace of $\mathbb{F}[z]$. Again, to check this, we need to verify the three conditions of Lemma 4.3.2.

Certainly the zero polynomial $p(z) = 0z^n + 0z^{n-1} + \cdots + 0z + 0$ is in U since $p(z)$ evaluated at 3 is 0. If $f(z), g(z) \in U$, then $f(3) = g(3) = 0$ so that $(f + g)(3) = f(3) + g(3) = 0 + 0 = 0$. Hence $f + g \in U$, which proves closure under addition. Similarly, $(af)(3) = af(3) = a0 = 0$ for any $a \in \mathbb{F}$, which proves closure under scalar multiplication.

Example 4.3.8. As in Example 4.1.6, let $D \subset \mathbb{R}$ be a subset of $\mathbb{R}$, and let $\mathcal{C}^\infty(D)$ denote the set of all smooth (a.k.a. continuously differentiable) functions with domain D and codomain $\mathbb{R}$. Then, under the same operations of point-wise addition and scalar multiplication, one can show that $\mathcal{C}^\infty(D)$ is a subspace of $\mathcal{C}(D)$.

Example 4.3.9. The subspaces of $\mathbb{R}^2$ consist of $\{0\}$, all lines through the origin, and $\mathbb{R}^2$ itself. The subspaces of $\mathbb{R}^3$ are $\{0\}$, all lines through the origin, all planes through the origin, and $\mathbb{R}^3$. In fact, these exhaust all subspaces of $\mathbb{R}^2$ and $\mathbb{R}^3$, respectively. To prove this, we will need further tools such as the notion of bases and dimensions to be discussed soon. In particular, this shows that lines and planes that do not pass through the origin are not subspaces (which is not so hard to show!).

Note that if U and U' are subspaces of V, then their intersection $U \cap U'$ is also a subspace (see Proof-writing Exercise 2 on page 38 and Figure 4.1). However, the union of two subspaces is not necessarily a subspace. Think, for example, of the union of two lines in $\mathbb{R}^2$, as in Figure 4.2.

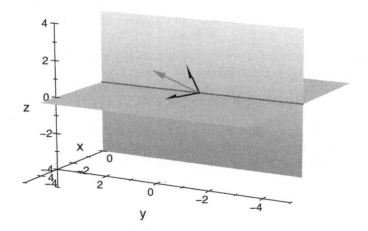

Fig. 4.1 The intersection $U \cap U'$ of two subspaces is a subspace

4.4 Sums and direct sums

Throughout this section, V is a vector space over $\mathbb{F}$, and $U_1, U_2 \subset V$ denote subspaces.

Definition 4.4.1. Let $U_1, U_2 \subset V$ be subspaces of V. Define the **(subspace) sum** of U_1 and U_2 to be the set

$$U_1 + U_2 = \{u_1 + u_2 \mid u_1 \in U_1, u_2 \in U_2\}.$$

Check as an exercise that $U_1 + U_2$ is a subspace of V. In fact, $U_1 + U_2$ is the smallest subspace of V that contains both U_1 and U_2.

Example 4.4.2. Let

$$U_1 = \{(x, 0, 0) \in \mathbb{F}^3 \mid x \in \mathbb{F}\},$$
$$U_2 = \{(0, y, 0) \in \mathbb{F}^3 \mid y \in \mathbb{F}\}.$$

Then

$$U_1 + U_2 = \{(x, y, 0) \in \mathbb{F}^3 \mid x, y \in \mathbb{F}\}. \tag{4.2}$$

If, alternatively, $U_2 = \{(y, y, 0) \in \mathbb{F}^3 \mid y \in \mathbb{F}\}$, then Equation (4.2) still holds.

If $U = U_1 + U_2$, then, for any $u \in U$, there exist $u_1 \in U_1$ and $u_2 \in U_2$ such that $u = u_1 + u_2$. If it so happens that u can be *uniquely* written as $u_1 + u_2$, then U is called the **direct sum** of U_1 and U_2.

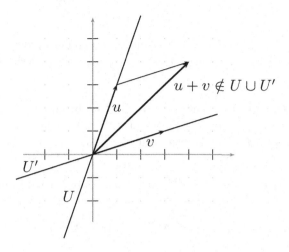

Fig. 4.2   The union $U \cup U'$ of two subspaces is not necessarily a subspace

Definition 4.4.3. Suppose every $u \in U$ can be uniquely written as $u = u_1 + u_2$ for $u_1 \in U_1$ and $u_2 \in U_2$. Then we use

$$U = U_1 \oplus U_2$$

to denote the **direct sum** of U_1 and U_2.

Example 4.4.4. Let

$$U_1 = \{(x, y, 0) \in \mathbb{R}^3 \mid x, y \in \mathbb{R}\},$$
$$U_2 = \{(0, 0, z) \in \mathbb{R}^3 \mid z \in \mathbb{R}\}.$$

Then $\mathbb{R}^3 = U_1 \oplus U_2$. However, if instead

$$U_2 = \{(0, w, z) \mid w, z \in \mathbb{R}\},$$

then $\mathbb{R}^3 = U_1 + U_2$ but is not the direct sum of U_1 and U_2.

Example 4.4.5. Let

$$U_1 = \{p \in \mathbb{F}[z] \mid p(z) = a_0 + a_2 z^2 + \cdots + a_{2m} z^{2m}\},$$
$$U_2 = \{p \in \mathbb{F}[z] \mid p(z) = a_1 z + a_3 z^3 + \cdots + a_{2m+1} z^{2m+1}\}.$$

Then $\mathbb{F}[z] = U_1 \oplus U_2$.

Proposition 4.4.6. *Let $U_1, U_2 \subset V$ be subspaces. Then $V = U_1 \oplus U_2$ if and only if the following two conditions hold:*

(1) $V = U_1 + U_2$;
(2) If $0 = u_1 + u_2$ with $u_1 \in U_1$ and $u_2 \in U_2$, then $u_1 = u_2 = 0$.

Proof.

("$\Longrightarrow$") Suppose $V = U_1 \oplus U_2$. Then Condition 1 holds by definition. Certainly $0 = 0 + 0$, and, since by uniqueness this is the only way to write $0 \in V$, we have $u_1 = u_2 = 0$.

("$\Longleftarrow$") Suppose Conditions 1 and 2 hold. By Condition 1, we have that, for all $v \in V$, there exist $u_1 \in U_1$ and $u_2 \in U_2$ such that $v = u_1 + u_2$. Suppose $v = w_1 + w_2$ with $w_1 \in U_1$ and $w_2 \in U_2$. Subtracting the two equations, we obtain

$$0 = (u_1 - w_1) + (u_2 - w_2),$$

where $u_1 - w_1 \in U_1$ and $u_2 - w_2 \in U_2$. By Condition 2, this implies $u_1 - w_1 = 0$ and $u_2 - w_2 = 0$, or equivalently $u_1 = w_1$ and $u_2 = w_2$, as desired. $\qquad\square$

Proposition 4.4.7. *Let* $U_1, U_2 \subset V$ *be subspaces. Then* $V = U_1 \oplus U_2$ *if and only if the following two conditions hold:*

(1) $V = U_1 + U_2$;
(2) $U_1 \cap U_2 = \{0\}$.

Proof.

("$\Longrightarrow$") Suppose $V = U_1 \oplus U_2$. Then Condition 1 holds by definition. If $u \in U_1 \cap U_2$, then $0 = u + (-u)$ with $u \in U_1$ and $-u \in U_2$ (why?). By Proposition 4.4.6, we have $u = 0$ and $-u = 0$ so that $U_1 \cap U_2 = \{0\}$.

("$\Longleftarrow$") Suppose Conditions 1 and 2 hold. To prove that $V = U_1 \oplus U_2$ holds, suppose that

$$0 = u_1 + u_2, \qquad \text{where } u_1 \in U_1 \text{ and } u_2 \in U_2. \tag{4.3}$$

By Proposition 4.4.6, it suffices to show that $u_1 = u_2 = 0$. Equation (4.3) implies that $u_1 = -u_2 \in U_2$. Hence $u_1 \in U_1 \cap U_2$, which in turn implies that $u_1 = 0$. It then follows that $u_2 = 0$ as well. $\qquad\square$

Everything in this section can be generalized to m subspaces $U_1, U_2, \ldots, U_m$, with the notable exception of Proposition 4.4.7. To see this consider the following example.

Example 4.4.8. Let

$$U_1 = \{(x, y, 0) \in \mathbb{F}^3 \mid x, y \in \mathbb{F}\},$$
$$U_2 = \{(0, 0, z) \in \mathbb{F}^3 \mid z \in \mathbb{F}\},$$
$$U_3 = \{(0, y, y) \in \mathbb{F}^3 \mid y \in \mathbb{F}\}.$$

Then certainly $\mathbb{F}^3 = U_1 + U_2 + U_3$, but $\mathbb{F}^3 \neq U_1 \oplus U_2 \oplus U_3$ since, for example,

$$(0, 0, 0) = (0, 1, 0) + (0, 0, 1) + (0, -1, -1).$$

But $U_1 \cap U_2 = U_1 \cap U_3 = U_2 \cap U_3 = \{0\}$ so that the analog of Proposition 4.4.7 does not hold.

Exercises for Chapter 4

Calculational Exercises

(1) For each of the following sets, either show that the set is a vector space or explain why it is not a vector space.

(a) The set $\mathbb{R}$ of real numbers under the usual operations of addition and multiplication.

(b) The set $\{(x, 0) \mid x \in \mathbb{R}\}$ under the usual operations of addition and multiplication on $\mathbb{R}^2$.

(c) The set $\{(x, 1) \mid x \in \mathbb{R}\}$ under the usual operations of addition and multiplication on $\mathbb{R}^2$.

(d) The set $\{(x, 0) \mid x \in \mathbb{R}, x \geq 0\}$ under the usual operations of addition and multiplication on $\mathbb{R}^2$.

(e) The set $\{(x, 1) \mid x \in \mathbb{R}, x \geq 0\}$ under the usual operations of addition and multiplication on $\mathbb{R}^2$.

(f) The set $\left\{ \begin{bmatrix} a & a+b \\ a+b & a \end{bmatrix} \mid a, b \in \mathbb{R} \right\}$ under the usual operations of addition and multiplication on $\mathbb{R}^{2 \times 2}$.

(g) The set $\left\{ \begin{bmatrix} a & a+b+1 \\ a+b & a \end{bmatrix} \mid a, b \in \mathbb{R} \right\}$ under the usual operations of addition and multiplication on $\mathbb{R}^{2 \times 2}$.

(2) Show that the space $V = \{(x_1, x_2, x_3) \in \mathbb{F}^3 \mid x_1 + 2x_2 + 2x_3 = 0\}$ forms a vector space.

(3) For each of the following sets, either show that the set is a subspace of $\mathcal{C}(\mathbb{R})$ or explain why it is not a subspace.

(a) The set $\{f \in \mathcal{C}(\mathbb{R}) \mid f(x) \leq 0, \ \forall x \in \mathbb{R}\}$.

(b) The set $\{f \in \mathcal{C}(\mathbb{R}) \mid f(0) = 0\}$.

(c) The set $\{f \in \mathcal{C}(\mathbb{R}) \mid f(0) = 2\}$.

(d) The set of all constant functions.

(e) The set $\{\alpha + \beta \sin(x) \mid \alpha, \beta \in \mathbb{R}\}$.

(4) Give an example of a nonempty subset $U \subset \mathbb{R}^2$ such that U is closed under scalar multiplication but is not a subspace of $\mathbb{R}^2$.

(5) Let $\mathbb{F}[z]$ denote the vector space of all polynomials with coefficients in $\mathbb{F}$, and define U to be the subspace of $\mathbb{F}[z]$ given by

$$U = \{az^2 + bz^5 \mid a, b \in \mathbb{F}\}.$$

Find a subspace W of $\mathbb{F}[z]$ such that $\mathbb{F}[z] = U \oplus W$.

Proof-Writing Exercises

(1) Let V be a vector space over $\mathbb{F}$. Then, given $a \in \mathbb{F}$ and $v \in V$ such that $av = 0$, prove that either $a = 0$ or $v = 0$.

(2) Let V be a vector space over $\mathbb{F}$, and suppose that W_1 and W_2 are subspaces of V. Prove that their intersection $W_1 \cap W_2$ is also a subspace of V.

(3) Prove or give a counterexample to the following claim:

Claim. *Let V be a vector space over $\mathbb{F}$, and suppose that W_1, W_2, and W_3 are subspaces of V such that $W_1 + W_3 = W_2 + W_3$. Then $W_1 = W_2$.*

(4) Prove or give a counterexample to the following claim:

Claim. *Let V be a vector space over $\mathbb{F}$, and suppose that W_1, W_2, and W_3 are subspaces of V such that $W_1 \oplus W_3 = W_2 \oplus W_3$. Then $W_1 = W_2$.*

Chapter 5

Span and Bases

The intuitive notion of dimension of a space as the number of coordinates one needs to uniquely specify a point in the space motivates the mathematical definition of dimension of a vector space. In this Chapter, we will first introduce the notions of linear span, linear independence, and basis of a vector space. Given a basis, we will find a bijective correspondence between coordinates and elements in a vector space, which leads to the definition of dimension of a vector space.

5.1 Linear span

As before, let V denote a vector space over $\mathbb{F}$. Given vectors $v_1, v_2, \ldots, v_m \in V$, a vector $v \in V$ is a **linear combination** of $(v_1, \ldots, v_m)$ if there exist scalars $a_1, \ldots, a_m \in \mathbb{F}$ such that

$$v = a_1 v_1 + a_2 v_2 + \cdots + a_m v_m.$$

Definition 5.1.1. The **linear span** (or simply **span**) of $(v_1, \ldots, v_m)$ is defined as

$$\mathrm{span}(v_1, \ldots, v_m) := \{a_1 v_1 + \cdots + a_m v_m \mid a_1, \ldots, a_m \in \mathbb{F}\}.$$

Lemma 5.1.2. *Let V be a vector space and $v_1, v_2, \ldots, v_m \in V$. Then*

(1) $v_j \in \mathrm{span}(v_1, v_2, \ldots, v_m)$.
(2) $\mathrm{span}(v_1, v_2, \ldots, v_m)$ is a subspace of V.
(3) If $U \subset V$ is a subspace such that $v_1, v_2, \ldots v_m \in U$, then $\mathrm{span}(v_1, v_2, \ldots, v_m) \subset U$.

Proof. Property 1 is obvious. For Property 2, note that $0 \in \mathrm{span}(v_1, v_2, \ldots, v_m)$ and that $\mathrm{span}(v_1, v_2, \ldots, v_m)$ is closed under addition and scalar multiplication. For Property 3, note that a subspace U of a vector space V is closed under addition and scalar multiplication. Hence, if $v_1, \ldots, v_m \in U$, then any linear combination $a_1 v_1 + \cdots + a_m v_m$ must also be an element of U. $\qquad\square$

Lemma 5.1.2 implies that $\mathrm{span}(v_1, v_2, \ldots, v_m)$ is the smallest subspace of V containing each of $v_1, v_2, \ldots, v_m$.

Definition 5.1.3. If $\mathrm{span}(v_1,\ldots,v_m) = V$, then we say that $(v_1,\ldots,v_m)$ **spans** V and we call V **finite-dimensional**. A vector space that is not finite-dimensional is called **infinite-dimensional**.

Example 5.1.4. The vectors $e_1 = (1,0,\ldots,0)$, $e_2 = (0,1,0,\ldots,0),\ldots,e_n = (0,\ldots,0,1)$ span $\mathbb{F}^n$. Hence $\mathbb{F}^n$ is finite-dimensional.

Example 5.1.5. The vectors $v_1 = (1,1,0)$ and $v_2 = (1,-1,0)$ span a subspace of $\mathbb{R}^3$. More precisely, if we write the vectors in $\mathbb{R}^3$ as 3-tuples of the form (x,y,z), then $\mathrm{span}(v_1,v_2)$ is the xy-plane in $\mathbb{R}^3$.

Example 5.1.6. Recall that if $p(z) = a_m z^m + a_{m-1} z^{m-1} + \cdots + a_1 z + a_0 \in \mathbb{F}[z]$ is a polynomial with coefficients in $\mathbb{F}$ such that $a_m \neq 0$, then we say that $p(z)$ has **degree** m. By convention, the degree of the zero polynomial $p(z) = 0$ is $-\infty$. We denote the degree of $p(z)$ by $\deg(p(z))$. Define

$$\mathbb{F}_m[z] = \text{set of all polynomials in } \mathbb{F}[z] \text{ of degree at most } m.$$

Then $\mathbb{F}_m[z] \subset \mathbb{F}[z]$ is a subspace since $\mathbb{F}_m[z]$ contains the zero polynomial and is closed under addition and scalar multiplication. In fact, $\mathbb{F}_m[z]$ is a finite-dimensional subspace of $\mathbb{F}[z]$ since

$$\mathbb{F}_m[z] = \mathrm{span}(1, z, z^2, \ldots, z^m).$$

At the same time, though, note that $\mathbb{F}[z]$ itself is infinite-dimensional. To see this, assume the contrary, namely that

$$\mathbb{F}[z] = \mathrm{span}(p_1(z),\ldots,p_k(z))$$

is spanned by a finite set of k polynomials $p_1(z),\ldots,p_k(z)$. Let $m = \max(\deg p_1(z),\ldots,\deg p_k(z))$. Then $z^{m+1} \in \mathbb{F}[z]$, but $z^{m+1} \notin \mathrm{span}(p_1(z),\ldots,p_k(z))$.

5.2 Linear independence

We are now going to define the notion of linear independence of a list of vectors. This concept will be extremely important in the sections that follow, and especially when we introduce bases and the dimension of a vector space.

Definition 5.2.1. A list of vectors $(v_1,\ldots,v_m)$ is called **linearly independent** if the only solution for $a_1,\ldots,a_m \in \mathbb{F}$ to the equation

$$a_1 v_1 + \cdots + a_m v_m = 0$$

is $a_1 = \cdots = a_m = 0$. In other words, the zero vector can only trivially be written as a linear combination of $(v_1,\ldots,v_m)$.

Definition 5.2.2. A list of vectors $(v_1, \ldots, v_m)$ is called **linearly dependent** if it is not linearly independent. That is, $(v_1, \ldots, v_m)$ is linear dependent if there exist $a_1, \ldots, a_m \in \mathbb{F}$, not all zero, such that

$$a_1 v_1 + \cdots + a_m v_m = 0.$$

Example 5.2.3. The vectors $(e_1, \ldots, e_m)$ of Example 5.1.4 are linearly independent. To see this, note that the only solution to the vector equation

$$0 = a_1 e_1 + \cdots + a_m e_m = (a_1, \ldots, a_m)$$

is $a_1 = \cdots = a_m = 0$. Alternatively, we can reinterpret this vector equation as the homogeneous linear system

$$\left. \begin{array}{r} a_1 \qquad\qquad = 0 \\ a_2 \qquad\quad = 0 \\ \ddots \quad \vdots \;\; \vdots \\ a_m = 0 \end{array} \right\},$$

which clearly has only the trivial solution. (See Section A.3.2 for the appropriate definitions.)

Example 5.2.4. The vectors $v_1 = (1, 1, 1), v_2 = (0, 1, -1)$, and $v_3 = (1, 2, 0)$ are linearly dependent. To see this, we need to consider the vector equation

$$\begin{aligned} a_1 v_1 + a_2 v_2 + a_3 v_3 &= a_1(1, 1, 1) + a_2(0, 1, -1) + a_3(1, 2, 0) \\ &= (a_1 + a_3, a_1 + a_2 + 2a_3, a_1 - a_2) = (0, 0, 0). \end{aligned}$$

Solving for a_1, a_2, and a_3, we see, for example, that $(a_1, a_2, a_3) = (1, 1, -1)$ is a non-zero solution. Alternatively, we can reinterpret this vector equation as the homogeneous linear system

$$\left. \begin{array}{r} a_1 \qquad + a_3 = 0 \\ a_1 + a_2 + 2a_3 = 0 \\ a_1 - a_2 \qquad = 0 \end{array} \right\}.$$

Using the techniques of Section A.3, we see that solving this linear system is equivalent to solving the following linear system:

$$\left. \begin{array}{r} a_1 \quad + a_3 = 0 \\ a_2 + a_3 = 0 \end{array} \right\}.$$

Note that this new linear system clearly has infinitely many solutions. In particular, the set of all solutions is given by

$$N = \{(a_1, a_2, a_3) \in \mathbb{F}^n \mid a_1 = a_2 = -a_3\} = \operatorname{span}((1, 1, -1)).$$

Example 5.2.5. The vectors $(1, z, \ldots, z^m)$ in the vector space $\mathbb{F}_m[z]$ are linearly independent. Requiring that

$$a_0 1 + a_1 z + \cdots + a_m z^m = 0$$

means that the polynomial on the left should be zero for all $z \in \mathbb{F}$. This is only possible for $a_0 = a_1 = \cdots = a_m = 0$.

An important consequence of the notion of linear independence is the fact that any vector in the span of a given list of linearly independent vectors can be uniquely written as a linear combination.

Lemma 5.2.6. *The list of vectors $(v_1, \ldots, v_m)$ is linearly independent if and only if every $v \in \mathrm{span}(v_1, \ldots, v_m)$ can be uniquely written as a linear combination of $(v_1, \ldots, v_m)$.*

Proof.

("$\Longrightarrow$") Assume that $(v_1, \ldots, v_m)$ is a linearly independent list of vectors. Suppose there are two ways of writing $v \in \mathrm{span}(v_1, \ldots, v_m)$ as a linear combination of the v_i:

$$v = a_1 v_1 + \cdots a_m v_m,$$
$$v = a_1' v_1 + \cdots a_m' v_m.$$

Subtracting the two equations yields $0 = (a_1 - a_1')v_1 + \cdots + (a_m - a_m')v_m$. Since $(v_1, \ldots, v_m)$ is linearly independent, the only solution to this equation is $a_1 - a_1' = 0, \ldots, a_m - a_m' = 0$, or equivalently $a_1 = a_1', \ldots, a_m = a_m'$.

("$\Longleftarrow$") Now assume that, for every $v \in \mathrm{span}(v_1, \ldots, v_m)$, there are unique $a_1, \ldots, a_m \in \mathbb{F}$ such that

$$v = a_1 v_1 + \cdots + a_m v_m.$$

This implies, in particular, that the only way the zero vector $v = 0$ can be written as a linear combination of $v_1, \ldots, v_m$ is with $a_1 = \cdots = a_m = 0$. This shows that $(v_1, \ldots, v_m)$ are linearly independent. $\qquad\square$

It is clear that if $(v_1, \ldots, v_m)$ is a list of linearly independent vectors, then the list $(v_1, \ldots, v_{m-1})$ is also linearly independent.

For the next lemma, we introduce the following notation: If we want to drop a vector v_j from a given list $(v_1, \ldots, v_m)$ of vectors, then we indicate the dropped vector by a hat. I.e., we write

$$(v_1, \ldots, \hat{v}_j, \ldots, v_m) = (v_1, \ldots, v_{j-1}, v_{j+1}, \ldots, v_m).$$

Lemma 5.2.7 (Linear Dependence Lemma). *If $(v_1, \ldots, v_m)$ is linearly dependent and $v_1 \neq 0$, then there exists an index $j \in \{2, \ldots, m\}$ such that the following two conditions hold.*

(1) $v_j \in \mathrm{span}(v_1, \ldots, v_{j-1})$.
(2) If v_j is removed from $(v_1, \ldots, v_m)$, then $\mathrm{span}(v_1, \ldots, \hat{v}_j, \ldots, v_m) = \mathrm{span}(v_1, \ldots, v_m)$.

Proof. Since $(v_1, \ldots, v_m)$ is linearly dependent there exist $a_1, \ldots, a_m \in \mathbb{F}$ not all zero such that $a_1 v_1 + \cdots + a_m v_m = 0$. Since by assumption $v_1 \neq 0$, not all of

$a_2, \ldots, a_m$ can be zero. Let $j \in \{2, \ldots, m\}$ be the largest index such that $a_j \neq 0$. Then we have

$$v_j = -\frac{a_1}{a_j} v_1 - \cdots - \frac{a_{j-1}}{a_j} v_{j-1}, \tag{5.1}$$

which implies Part 1.

Let $v \in \text{span}(v_1, \ldots, v_m)$. This means, by definition, that there exist scalars $b_1, \ldots, b_m \in \mathbb{F}$ such that

$$v = b_1 v_1 + \cdots + b_m v_m.$$

The vector v_j that we determined in Part 1 can be replaced by Equation (5.1) so that v is written as a linear combination of $(v_1, \ldots, \hat{v}_j, \ldots, v_m)$. Hence, $\text{span}(v_1, \ldots, \hat{v}_j, \ldots, v_m) = \text{span}(v_1, \ldots, v_m)$. □

Example 5.2.8. The list $(v_1, v_2, v_3) = ((1,1),(1,2),(1,0))$ of vectors spans $\mathbb{R}^2$. To see this, take any vector $v = (x,y) \in \mathbb{R}^2$. We want to show that v can be written as a linear combination of $(1,1),(1,2),(1,0)$, i.e., that there exist scalars $a_1, a_2, a_3 \in \mathbb{F}$ such that

$$v = a_1(1,1) + a_2(1,2) + a_3(1,0),$$

or equivalently that

$$(x,y) = (a_1 + a_2 + a_3, a_1 + 2a_2).$$

Clearly $a_1 = y$, $a_2 = 0$, and $a_3 = x - y$ form a solution for any choice of $x, y \in \mathbb{R}$, and so $\mathbb{R}^2 = \text{span}((1,1),(1,2),(1,0))$. However, note that

$$2(1,1) - (1,2) - (1,0) = (0,0), \tag{5.2}$$

which shows that the list $((1,1),(1,2),(1,0))$ is linearly dependent. The Linear Dependence Lemma 5.2.7 thus states that one of the vectors can be dropped from $((1,1),(1,2),(1,0))$ and that the resulting list of vectors will still span $\mathbb{R}^2$. Indeed, by Equation (5.2),

$$v_3 = (1,0) = 2(1,1) - (1,2) = 2v_1 - v_2,$$

and so $\text{span}((1,1),(1,2),(1,0)) = \text{span}((1,1),(1,2))$.

The next result shows that linearly independent lists of vectors that span a finite-dimensional vector space are the smallest possible spanning sets.

Theorem 5.2.9. *Let V be a finite-dimensional vector space. Suppose that $(v_1, \ldots, v_m)$ is a linearly independent list of vectors that spans V, and let $(w_1, \ldots, w_n)$ be any list that spans V. Then $m \leq n$.*

Proof. The proof uses the following iterative procedure: start with an arbitrary list of vectors $\mathcal{S}_0 = (w_1, \ldots, w_n)$ such that $V = \text{span}(\mathcal{S}_0)$. At the kth step of the procedure, we construct a new list $\mathcal{S}_k$ by replacing some vector w_{j_k} by the vector v_k such that $\mathcal{S}_k$ still spans V. Repeating this for all v_k then produces a new list $\mathcal{S}_m$

of length n that contains each of $v_1, \ldots, v_m$, which then proves that $m \leq n$. Let us now discuss each step in this procedure in detail.

Step 1. Since $(w_1, \ldots, w_n)$ spans V, adding a new vector to the list makes the new list linearly dependent. Hence $(v_1, w_1, \ldots, w_n)$ is linearly dependent. By Lemma 5.2.7, there exists an index j_1 such that

$$w_{j_1} \in \text{span}(v_1, w_1, \ldots, w_{j_1-1}).$$

Hence $\mathcal{S}_1 = (v_1, w_1, \ldots, \hat{w}_{j_1}, \ldots, w_n)$ spans V. In this step, we added the vector v_1 and removed the vector w_{j_1} from $\mathcal{S}_0$.

Step k. Suppose that we already added $v_1, \ldots, v_{k-1}$ to our spanning list and removed the vectors $w_{j_1}, \ldots, w_{j_{k-1}}$. It is impossible that we have reached the situation where all of the vectors $w_1, \ldots, w_n$ have been removed from the spanning list at this step if $k \leq m$ because then we would have $V = \text{span}(v_1, \ldots, v_{k-1})$ which would allow v_k to be expressed as a linear combination of $v_1, \ldots, v_{k-1}$ (in contradiction with the assumption of linear independence of $v_1, \ldots, v_n$).

Now, call the list reached at this step $\mathcal{S}_{k-1}$, and note that $V = \text{span}(\mathcal{S}_{k-1})$. Add the vector v_k to $\mathcal{S}_{k-1}$. By the same arguments as before, adjoining the extra vector v_k to the spanning list $\mathcal{S}_{k-1}$ yields a list of linearly dependent vectors. Hence, by Lemma 5.2.7, there exists an index j_k such that $\mathcal{S}_{k-1}$ with v_k added and w_{j_k} removed still spans V. The fact that $(v_1, \ldots, v_k)$ is linearly independent ensures that the vector removed is indeed among the w_j. Call the new list $\mathcal{S}_k$, and note that $V = \text{span}(\mathcal{S}_k)$.

The final list $\mathcal{S}_m$ is $\mathcal{S}_0$ but with each $v_1, \ldots, v_m$ added and each $w_{j_1}, \ldots, w_{j_m}$ removed. Moreover, note that $\mathcal{S}_m$ has length n and still spans V. It follows that $m \leq n$. $\qquad\square$

5.3   Bases

A basis of a finite-dimensional vector space is a spanning list that is also linearly independent. We will see that all bases for finite-dimensional vector spaces have the same length. This length will then be called the **dimension** of our vector space.

Definition 5.3.1. A list of vectors $(v_1, \ldots, v_m)$ is a **basis** for the finite-dimensional vector space V if $(v_1, \ldots, v_m)$ is linearly independent and $V = \text{span}(v_1, \ldots, v_m)$.

If $(v_1, \ldots, v_m)$ forms a basis of V, then, by Lemma 5.2.6, every vector $v \in V$ can be uniquely written as a linear combination of $(v_1, \ldots, v_m)$.

Example 5.3.2. $(e_1, \ldots, e_n)$ is a basis of $\mathbb{F}^n$. There are, of course, other bases. For example, $((1,2), (1,1))$ is a basis of $\mathbb{F}^2$. Note that the list $((1,1))$ is also linearly independent, but it does not span $\mathbb{F}^2$ and hence is not a basis.

Example 5.3.3. $(1, z, z^2, \ldots, z^m)$ is a basis of $\mathbb{F}_m[z]$.

Theorem 5.3.4 (Basis Reduction Theorem). *If $V = \mathrm{span}(v_1, \ldots, v_m)$, then either $(v_1, \ldots, v_m)$ is a basis of V or some v_i can be removed to obtain a basis of V.*

Proof. Suppose $V = \mathrm{span}(v_1, \ldots, v_m)$. We start with the list $\mathcal{S} = (v_1, \ldots, v_m)$ and sequentially run through all vectors v_k for $k = 1, 2, \ldots, m$ to determine whether to keep or remove them from $\mathcal{S}$:

Step 1. If $v_1 = 0$, then remove v_1 from $\mathcal{S}$. Otherwise, leave $\mathcal{S}$ unchanged.

Step k. If $v_k \in \mathrm{span}(v_1, \ldots, v_{k-1})$, then remove v_k from $\mathcal{S}$. Otherwise, leave $\mathcal{S}$ unchanged.

The final list $\mathcal{S}$ still spans V since, at each step, a vector was only discarded if it was already in the span of the previous vectors. The process also ensures that no vector is in the span of the previous vectors. Hence, by the Linear Dependence Lemma 5.2.7, the final list $\mathcal{S}$ is linearly independent. It follows that $\mathcal{S}$ is a basis of V. $\qquad\square$

Example 5.3.5. To see how Basis Reduction Theorem 5.3.4 works, consider the list of vectors

$$\mathcal{S} = ((1, -1, 0), (2, -2, 0), (-1, 0, 1), (0, -1, 1), (0, 1, 0)).$$

This list does not form a basis for $\mathbb{R}^3$ as it is not linearly independent. However, it is clear that $\mathbb{R}^3 = \mathrm{span}(\mathcal{S})$ since any arbitrary vector $v = (x, y, z) \in \mathbb{R}^3$ can be written as the following linear combination over $\mathcal{S}$:

$$v = (x + z)(1, -1, 0) + 0(2, -2, 0) + (z)(-1, 0, 1) + 0(0, -1, 1) + (x + y + z)(0, 1, 0).$$

In fact, since the coefficients of $(2, -2, 0)$ and $(0, -1, 1)$ in this linear combination are both zero, it suggests that they add nothing to the span of the subset

$$\mathcal{B} = ((1, -1, 0), (-1, 0, 1), (0, 1, 0))$$

of $\mathcal{S}$. Moreover, one can show that $\mathcal{B}$ is a basis for $\mathbb{R}^3$, and it is exactly the basis produced by applying the process from the proof of Theorem 5.3.4 (as you should be able to verify).

Corollary 5.3.6. *Every finite-dimensional vector space has a basis.*

Proof. By definition, a finite-dimensional vector space has a spanning list. By the Basis Reduction Theorem 5.3.4, any spanning list can be reduced to a basis. $\qquad\square$

Theorem 5.3.7 (Basis Extension Theorem). *Every linearly independent list of vectors in a finite-dimensional vector space V can be extended to a basis of V.*

Proof. Suppose V is finite-dimensional and that $(v_1, \ldots, v_m)$ is linearly independent. Since V is finite-dimensional, there exists a list $(w_1, \ldots, w_n)$ of vectors that spans V. We wish to adjoin some of the w_k to $(v_1, \ldots, v_m)$ in order to create a basis of V.

Step 1. If $w_1 \in \mathrm{span}(v_1, \ldots, v_m)$, then let $\mathcal{S} = (v_1, \ldots, v_m)$. Otherwise, $\mathcal{S} = (v_1, \ldots, v_m, w_1)$.

Step k. If $w_k \in \mathrm{span}(\mathcal{S})$, then leave $\mathcal{S}$ unchanged. Otherwise, adjoin w_k to $\mathcal{S}$.

After each step, the list $\mathcal{S}$ is still linearly independent since we only adjoined w_k if w_k was not in the span of the previous vectors. After n steps, $w_k \in \mathrm{span}(\mathcal{S})$ for all $k = 1, 2, \ldots, n$. Since $(w_1, \ldots, w_n)$ was a spanning list, $\mathcal{S}$ spans V so that $\mathcal{S}$ is indeed a basis of V. $\qquad\qquad\square$

Example 5.3.8. Take the two vectors $v_1 = (1, 1, 0, 0)$ and $v_2 = (1, 0, 1, 0)$ in $\mathbb{R}^4$. One may easily check that these two vectors are linearly independent, but they do not form a basis of $\mathbb{R}^4$. We know that (e_1, e_2, e_3, e_4) spans $\mathbb{R}^4$. (In fact, it is even a basis.) Following the algorithm outlined in the proof of the Basis Extension Theorem, we see that $e_1 \notin \mathrm{span}(v_1, v_2)$. Hence, we adjoin e_1 to obtain $\mathcal{S} = (v_1, v_2, e_1)$. Note that now

$$e_2 = (0, 1, 0, 0) = 1v_1 + 0v_2 + (-1)e_1$$

so that $e_2 \in \mathrm{span}(v_1, v_2, e_1)$, and so we leave $\mathcal{S}$ unchanged. Similarly,

$$e_3 = (0, 0, 1, 0) = 0v_1 + 1v_2 + (-1)e_1,$$

and hence $e_3 \in \mathrm{span}(v_1, v_2, e_1)$, which means that we again leave $\mathcal{S}$ unchanged. Finally, $e_4 \notin \mathrm{span}(v_1, v_2, e_1)$, and so we adjoin it to obtain a basis (v_1, v_2, e_1, e_4) of $\mathbb{R}^4$.

5.4   Dimension

We now come to the important definition of the dimension of a finite-dimensional vector space, which corresponds to the intuitive notion that $\mathbb{R}^2$ has dimension 2, $\mathbb{R}^3$ has dimension 3, and, more generally, that $\mathbb{R}^n$ has dimension n. This is precisely the length of every basis for each of these vector spaces, which prompts the following definition.

Definition 5.4.1. We call the length of any basis for V (which is well-defined by Theorem 5.4.2 below) the **dimension** of V, and we denote this by $\dim(V)$.

Note that Definition 5.4.1 only makes sense if, in fact, every basis for a given finite-dimensional vector space has the same length. This is true by the following theorem.

Theorem 5.4.2. *Let V be a finite-dimensional vector space. Then any two bases of V have the same length.*

Proof. Let $(v_1, \ldots, v_m)$ and $(w_1, \ldots, w_n)$ be two bases of V. Both span V. By Theorem 5.2.9, we have $m \leq n$ since $(v_1, \ldots, v_m)$ is linearly independent. By the same theorem, we also have $n \leq m$ since $(w_1, \ldots, w_n)$ is linearly independent. Hence $n = m$, as asserted. □

Example 5.4.3. $\dim(\mathbb{F}^n) = n$ and $\dim(\mathbb{F}_m[z]) = m + 1$. Note that $\dim(\mathbb{C}^n) = n$ as a complex vector space, whereas $\dim(\mathbb{C}^n) = 2n$ as a real vector space. This comes from the fact that we can view $\mathbb{C}$ itself as a real vector space of dimension 2 with basis $(1, i)$.

Theorem 5.4.4. *Let V be a finite-dimensional vector space with $\dim(V) = n$. Then:*

(1) If $U \subset V$ is a subspace of V, then $\dim(U) \leq \dim(V)$.
(2) If $V = \mathrm{span}(v_1, \ldots, v_n)$, then $(v_1, \ldots, v_n)$ is a basis of V.
(3) If $(v_1, \ldots, v_n)$ is linearly independent in V, then $(v_1, \ldots, v_n)$ is a basis of V.

Point 1 implies, in particular, that every subspace of a finite-dimensional vector space is finite-dimensional. Points 2 and 3 show that if the dimension of a vector space is known to be n, then, to check that a list of n vectors is a basis, it is enough to check whether it spans V (resp. is linearly independent).

Proof. To prove Point 1, first note that U is necessarily finite-dimensional (otherwise we could find a list of linearly independent vectors longer than $\dim(V)$). Therefore, by Corollary 5.3.6, U has a basis, $(u_1, \ldots, u_m)$, say. This list is linearly independent in both U and V. By the Basis Extension Theorem 5.3.7, we can extend $(u_1, \ldots, u_m)$ to a basis for V, which is of length n since $\dim(V) = n$. This implies that $m \leq n$, as desired.

To prove Point 2, suppose that $(v_1, \ldots, v_n)$ spans V. Then, by the Basis Reduction Theorem 5.3.4, this list can be reduced to a basis. However, every basis of V has length n; hence, no vector needs to be removed from $(v_1, \ldots, v_n)$. It follows that $(v_1, \ldots, v_n)$ is already a basis of V.

Point 3 is proven in a similar fashion. Suppose $(v_1, \ldots, v_n)$ is linearly independent. By the Basis Extension Theorem 5.3.7, this list can be extended to a basis. However, every basis has length n; hence, no vector needs to be added to $(v_1, \ldots, v_n)$. It follows that $(v_1, \ldots, v_n)$ is already a basis of V. □

We conclude this chapter with some additional interesting results on bases and dimensions. The first one combines the concepts of basis and direct sum.

Theorem 5.4.5. *Let $U \subset V$ be a subspace of a finite-dimensional vector space V. Then there exists a subspace $W \subset V$ such that $V = U \oplus W$.*

Proof. Let $(u_1, \ldots, u_m)$ be a basis of U. By Theorem 5.4.4(1), we know that $m \leq \dim(V)$. Hence, by the Basis Extension Theorem 5.3.7, $(u_1, \ldots, u_m)$ can be extended to a basis $(u_1, \ldots, u_m, w_1, \ldots, w_n)$ of V. Let $W = \text{span}(w_1, \ldots, w_n)$.

To show that $V = U \oplus W$, we need to show that $V = U + W$ and $U \cap W = \{0\}$. Since $V = \text{span}(u_1, \ldots, u_m, w_1, \ldots, w_n)$ where $(u_1, \ldots, u_m)$ spans U and $(w_1, \ldots, w_n)$ spans W, it is clear that $V = U + W$.

To show that $U \cap W = \{0\}$, let $v \in U \cap W$. Then there exist scalars $a_1, \ldots, a_m, b_1, \ldots, b_n \in \mathbb{F}$ such that

$$v = a_1 u_1 + \cdots + a_m u_m = b_1 w_1 + \cdots + b_n w_n,$$

or equivalently that

$$a_1 u_1 + \cdots + a_m u_m - b_1 w_1 - \cdots - b_n w_n = 0.$$

Since $(u_1, \ldots, u_m, w_1, \ldots, w_n)$ forms a basis of V and hence is linearly independent, the only solution to this equation is $a_1 = \cdots = a_m = b_1 = \cdots = b_n = 0$. Hence $v = 0$, proving that indeed $U \cap W = \{0\}$. $\qquad\square$

Theorem 5.4.6. *If $U, W \subset V$ are subspaces of a finite-dimensional vector space, then*

$$\dim(U + W) = \dim(U) + \dim(W) - \dim(U \cap W).$$

Proof. Let $(v_1, \ldots, v_n)$ be a basis of $U \cap W$. By the Basis Extension Theorem 5.3.7, there exist $(u_1, \ldots, u_k)$ and $(w_1, \ldots, w_\ell)$ such that $(v_1, \ldots, v_n, u_1, \ldots, u_k)$ is a basis of U and $(v_1, \ldots, v_n, w_1, \ldots, w_\ell)$ is a basis of W. It suffices to show that

$$\mathcal{B} = (v_1, \ldots, v_n, u_1, \ldots, u_k, w_1, \ldots, w_\ell)$$

is a basis of $U + W$ since then

$$\dim(U + W) = n + k + \ell = (n + k) + (n + \ell) - n = \dim(U) + \dim(W) - \dim(U \cap W).$$

Clearly $\text{span}(v_1, \ldots, v_n, u_1, \ldots, u_k, w_1, \ldots, w_\ell)$ contains U and W, and hence $U + W$. To show that $\mathcal{B}$ is a basis, it remains to show that $\mathcal{B}$ is linearly independent. Suppose

$$a_1 v_1 + \cdots + a_n v_n + b_1 u_1 + \cdots + b_k u_k + c_1 w_1 + \cdots + c_\ell w_\ell = 0, \qquad (5.3)$$

and let $u = a_1 v_1 + \cdots + a_n v_n + b_1 u_1 + \cdots + b_k u_k \in U$. Then, by Equation (5.3), we also have that $u = -c_1 w_1 - \cdots - c_\ell w_\ell \in W$, which implies that $u \in U \cap W$. Hence, there exist scalars $a_1', \ldots, a_n' \in \mathbb{F}$ such that $u = a_1' v_1 + \cdots + a_n' v_n$. Since there is a unique linear combination of the linearly independent vectors $(v_1, \ldots, v_n, u_1, \ldots, u_k)$ that describes u, we must have $b_1 = \cdots = b_k = 0$ and $a_1 = a_1', \ldots, a_n = a_n'$. Since $(v_1, \ldots, v_n, w_1, \ldots, w_\ell)$ is also linearly independent, it further follows that $a_1 = \cdots = a_n = c_1 = \cdots = c_\ell = 0$. Hence, Equation (5.3) only has the trivial solution, which implies that $\mathcal{B}$ is a basis. $\qquad\square$

Exercises for Chapter 5

Calculational Exercises

(1) Show that the vectors $v_1 = (1,1,1)$, $v_2 = (1,2,3)$, and $v_3 = (2,-1,1)$ are linearly independent in $\mathbb{R}^3$. Write $v = (1,-2,5)$ as a linear combination of v_1, v_2, and v_3.

(2) Consider the complex vector space $V = \mathbb{C}^3$ and the list (v_1, v_2, v_3) of vectors in V, where

$$v_1 = (i,0,0), \quad v_2 = (i,1,0), \quad v_3 = (i,i,-1).$$

    (a) Prove that $\mathrm{span}(v_1, v_2, v_3) = V$.
    (b) Prove or disprove: (v_1, v_2, v_3) is a basis for V.

(3) Determine the dimension of each of the following subspaces of $\mathbb{F}^4$.

    (a) $\{(x_1, x_2, x_3, x_4) \in \mathbb{F}^4 \mid x_4 = 0\}$.
    (b) $\{(x_1, x_2, x_3, x_4) \in \mathbb{F}^4 \mid x_4 = x_1 + x_2\}$.
    (c) $\{(x_1, x_2, x_3, x_4) \in \mathbb{F}^4 \mid x_4 = x_1 + x_2, x_3 = x_1 - x_2\}$.
    (d) $\{(x_1, x_2, x_3, x_4) \in \mathbb{F}^4 \mid x_4 = x_1 + x_2, x_3 = x_1 - x_2, x_3 + x_4 = 2x_1\}$.
    (e) $\{(x_1, x_2, x_3, x_4) \in \mathbb{F}^4 \mid x_1 = x_2 = x_3 = x_4\}$.

(4) Determine the value of $\lambda \in \mathbb{R}$ for which each list of vectors is linear dependent.

    (a) $((\lambda, -1, -1), (-1, \lambda, -1), (-1, -1, \lambda))$ as a subset of $\mathbb{R}^3$.
    (b) $\left(\sin^2(x), \cos(2x), \lambda\right)$ as a subset of $C(\mathbb{R})$.

(5) Consider the real vector space $V = \mathbb{R}^4$. For each of the following five statements, provide either a proof or a counterexample.

    (a) $\dim V = 4$.
    (b) $\mathrm{span}((1,1,0,0), (0,1,1,0), (0,0,1,1)) = V$.
    (c) The list $((1,-1,0,0), (0,1,-1,0), (0,0,1,-1), (-1,0,0,1))$ is linearly independent.
    (d) Every list of four vectors $v_1, \ldots, v_4 \in V$, such that $\mathrm{span}(v_1, \ldots, v_4) = V$, is linearly independent.
    (e) Let v_1 and v_2 be two linearly independent vectors in V. Then, there exist vectors $u, w \in V$, such that (v_1, v_2, u, w) is a basis for V.

Proof-Writing Exercises

(1) Let V be a vector space over $\mathbb{F}$ and define $U = \mathrm{span}(u_1, u_2, \ldots, u_n)$, where for each $i = 1, \ldots, n$, $u_i \in V$. Now suppose $v \in U$. Prove

$$U = \mathrm{span}(v, u_1, u_2, \ldots, u_n).$$

(2) Let V be a vector space over $\mathbb{F}$, and suppose that the list $(v_1, v_2, \ldots, v_n)$ of vectors spans V, where each $v_i \in V$. Prove that the list

$$(v_1 - v_2, v_2 - v_3, v_3 - v_4, \ldots, v_{n-2} - v_{n-1}, v_{n-1} - v_n, v_n)$$

also spans V.

(3) Let V be a vector space over $\mathbb{F}$, and suppose that $(v_1, v_2, \ldots, v_n)$ is a linearly independent list of vectors in V. Given any $w \in V$ such that

$$(v_1 + w, v_2 + w, \ldots, v_n + w)$$

is a linearly dependent list of vectors in V, prove that $w \in \operatorname{span}(v_1, v_2, \ldots, v_n)$.

(4) Let V be a finite-dimensional vector space over $\mathbb{F}$ with $\dim(V) = n$ for some $n \in \mathbb{Z}_+$. Prove that there are n one-dimensional subspaces $U_1, U_2, \ldots, U_n$ of V such that

$$V = U_1 \oplus U_2 \oplus \cdots \oplus U_n.$$

(5) Let V be a finite-dimensional vector space over $\mathbb{F}$, and suppose that U is a subspace of V for which $\dim(U) = \dim(V)$. Prove that $U = V$.

(6) Let $\mathbb{F}_m[z]$ denote the vector space of all polynomials with degree less than or equal to $m \in \mathbb{Z}_+$ and having coefficient over $\mathbb{F}$, and suppose that $p_0, p_1, \ldots, p_m \in \mathbb{F}_m[z]$ satisfy $p_j(2) = 0$. Prove that $(p_0, p_1, \ldots, p_m)$ is a linearly dependent list of vectors in $\mathbb{F}_m[z]$.

(7) Let U and V be five-dimensional subspaces of $\mathbb{R}^9$. Prove that $U \cap V \neq \{0\}$.

(8) Let V be a finite-dimensional vector space over $\mathbb{F}$, and suppose that $U_1, U_2, \ldots, U_m$ are any m subspaces of V. Prove that

$$\dim(U_1 + U_2 + \cdots + U_m) \leq \dim(U_1) + \dim(U_2) + \cdots + \dim(U_m).$$

Chapter 6

Linear Maps

As discussed in Chapter 1, one of the main goals of Linear Algebra is the characterization of solutions to a system of m linear equations in n unknowns $x_1, \ldots, x_n$,

$$\left. \begin{array}{ccc} a_{11}x_1 + \cdots + a_{1n}x_n = b_1 \\ \vdots \quad \vdots \quad \vdots \\ a_{m1}x_1 + \cdots + a_{mn}x_n = b_m \end{array} \right\},$$

where each of the coefficients a_{ij} and b_i is in $\mathbb{F}$. Linear maps and their properties give us insight into the characteristics of solutions to linear systems.

6.1 Definition and elementary properties

Throughout this chapter, V and W denote vector spaces over $\mathbb{F}$. We are going to study functions from V into W that have the special properties given in the following definition.

Definition 6.1.1. A function $T : V \to W$ is called **linear** if

$$T(u + v) = T(u) + T(v), \qquad \text{for all } u, v \in V, \tag{6.1}$$
$$T(av) = aT(v), \qquad \text{for all } a \in \mathbb{F} \text{ and } v \in V. \tag{6.2}$$

The set of all linear maps from V to W is denoted by $\mathcal{L}(V, W)$. We sometimes write Tv for $T(v)$. Linear maps are also called **linear transformations**.

Moreover, if $V = W$, then we write $\mathcal{L}(V, V) = \mathcal{L}(V)$ and call $T \in \mathcal{L}(V)$ a **linear operator** on V.

Example 6.1.2.

(1) The **zero map** $0 : V \to W$ mapping every element $v \in V$ to $0 \in W$ is linear.
(2) The **identity map** $I : V \to V$ defined as $Iv = v$ is linear.
(3) Let $T : \mathbb{F}[z] \to \mathbb{F}[z]$ be the **differentiation map** defined as $Tp(z) = p'(z)$. Then, for two polynomials $p(z), q(z) \in \mathbb{F}[z]$, we have

$$T(p(z) + q(z)) = (p(z) + q(z))' = p'(z) + q'(z) = T(p(z)) + T(q(z)).$$

Similarly, for a polynomial $p(z) \in \mathbb{F}[z]$ and a scalar $a \in \mathbb{F}$, we have

$$T(ap(z)) = (ap(z))' = ap'(z) = aT(p(z)).$$

Hence T is linear.

(4) Let $T : \mathbb{R}^2 \to \mathbb{R}^2$ be the map given by $T(x, y) = (x - 2y, 3x + y)$. Then, for $(x, y), (x', y') \in \mathbb{R}^2$, we have

$$T((x, y) + (x', y')) = T(x + x', y + y') = (x + x' - 2(y + y'), 3(x + x') + y + y')$$
$$= (x - 2y, 3x + y) + (x' - 2y', 3x' + y') = T(x, y) + T(x', y').$$

Similarly, for $(x, y) \in \mathbb{R}^2$ and $a \in \mathbb{F}$, we have

$$T(a(x, y)) = T(ax, ay) = (ax - 2ay, 3ax + ay) = a(x - 2y, 3x + y) = aT(x, y).$$

Hence T is linear. More generally, any map $T : \mathbb{F}^n \to \mathbb{F}^m$ defined by

$$T(x_1, \ldots, x_n) = (a_{11}x_1 + \cdots + a_{1n}x_n, \ldots, a_{m1}x_1 + \cdots + a_{mn}x_n)$$

with $a_{ij} \in \mathbb{F}$ is linear.

(5) Not all functions are linear! For example, the exponential function $f(x) = e^x$ is not linear since $e^{2x} \neq 2e^x$ in general. Also, the function $f : \mathbb{F} \to \mathbb{F}$ given by $f(x) = x - 1$ is not linear since $f(x + y) = (x + y) - 1 \neq (x - 1) + (y - 1) = f(x) + f(y)$.

An important result is that linear maps are already completely determined if their values on basis vectors are specified.

Theorem 6.1.3. *Let $(v_1, \ldots, v_n)$ be a basis of V and $(w_1, \ldots, w_n)$ be an arbitrary list of vectors in W. Then there exists a unique linear map*

$$T : V \to W \quad \text{such that } T(v_i) = w_i, \, \forall \, i = 1, 2, \ldots, n.$$

Proof. First we verify that there is at most one linear map T with $T(v_i) = w_i$. Take any $v \in V$. Since $(v_1, \ldots, v_n)$ is a basis of V, there are unique scalars $a_1, \ldots, a_n \in \mathbb{F}$ such that $v = a_1 v_1 + \cdots + a_n v_n$. By linearity, we have

$$T(v) = T(a_1 v_1 + \cdots + a_n v_n) = a_1 T(v_1) + \cdots + a_n T(v_n) = a_1 w_1 + \cdots + a_n w_n, \quad (6.3)$$

and hence $T(v)$ is completely determined. To show existence, use Equation (6.3) to define T. It remains to show that this T is linear and that $T(v_i) = w_i$. These two conditions are not hard to show and are left to the reader. $\square$

The set of linear maps $\mathcal{L}(V, W)$ is itself a vector space. For $S, T \in \mathcal{L}(V, W)$ addition is defined as

$$(S + T)v = Sv + Tv, \quad \text{for all } v \in V.$$

For $a \in \mathbb{F}$ and $T \in \mathcal{L}(V, W)$, scalar multiplication is defined as

$$(aT)(v) = a(Tv), \quad \text{for all } v \in V.$$

You should verify that $S+T$ and aT are indeed linear maps and that all properties of a vector space are satisfied.

In addition to the operations of vector addition and scalar multiplication, we can also define the **composition of linear maps**. Let V, U, W be vector spaces over $\mathbb{F}$. Then, for $S \in \mathcal{L}(U, V)$ and $T \in \mathcal{L}(V, W)$, we define $T \circ S \in \mathcal{L}(U, W)$ by

$$(T \circ S)(u) = T(S(u)), \quad \text{for all } u \in U.$$

The map $T \circ S$ is often also called the **product** of T and S denoted by TS. It has the following properties:

(1) **Associativity:** $(T_1 T_2)T_3 = T_1(T_2 T_3)$, for all $T_1 \in \mathcal{L}(V_1, V_0)$, $T_2 \in \mathcal{L}(V_2, V_1)$ and $T_3 \in \mathcal{L}(V_3, V_2)$.
(2) **Identity:** $TI = IT = T$, where $T \in \mathcal{L}(V, W)$ and where I in TI is the identity map in $\mathcal{L}(V, V)$ whereas the I in IT is the identity map in $\mathcal{L}(W, W)$.
(3) **Distributivity:** $(T_1 + T_2)S = T_1 S + T_2 S$ and $T(S_1 + S_2) = TS_1 + TS_2$, where $S, S_1, S_2 \in \mathcal{L}(U, V)$ and $T, T_1, T_2 \in \mathcal{L}(V, W)$.

Note that the product of linear maps is not always commutative. For example, if we take $T \in \mathcal{L}(\mathbb{F}[z], \mathbb{F}[z])$ to be the differentiation map $Tp(z) = p'(z)$ and $S \in \mathcal{L}(\mathbb{F}[z], \mathbb{F}[z])$ to be the map $Sp(z) = z^2 p(z)$, then

$$(ST)p(z) = z^2 p'(z) \quad \text{but} \quad (TS)p(z) = z^2 p'(z) + 2zp(z).$$

6.2 Null spaces

Definition 6.2.1. Let $T : V \to W$ be a linear map. Then the **null space** (a.k.a. **kernel**) of T is the set of all vectors in V that are mapped to zero by T. I.e.,

$$\text{null}\,(T) = \{v \in V \mid Tv = 0\}.$$

Example 6.2.2. Let $T \in \mathcal{L}(\mathbb{F}[z], \mathbb{F}[z])$ be the differentiation map $Tp(z) = p'(z)$. Then

$$\text{null}\,(T) = \{p \in \mathbb{F}[z] \mid p(z) \text{ is constant}\}.$$

Example 6.2.3. Consider the linear map $T(x, y) = (x - 2y, 3x + y)$ of Example 6.1.2. To determine the null space, we need to solve $T(x, y) = (0, 0)$, which is equivalent to the system of linear equations

$$\left.\begin{array}{r} x - 2y = 0 \\ 3x + y = 0 \end{array}\right\}.$$

We see that the only solution is $(x, y) = (0, 0)$ so that $\text{null}\,(T) = \{(0, 0)\}$.

Proposition 6.2.4. *Let $T : V \to W$ be a linear map. Then $\text{null}\,(T)$ is a subspace of V.*

Proof. We need to show that $0 \in \text{null}\,(T)$ and that $\text{null}\,(T)$ is closed under addition and scalar multiplication. By linearity, we have

$$T(0) = T(0 + 0) = T(0) + T(0)$$

so that $T(0) = 0$. Hence $0 \in \text{null}\,(T)$. For closure under addition, let $u, v \in \text{null}\,(T)$. Then

$$T(u + v) = T(u) + T(v) = 0 + 0 = 0,$$

and hence $u + v \in \text{null}\,(T)$. Similarly, for closure under scalar multiplication, let $u \in \text{null}\,(T)$ and $a \in \mathbb{F}$. Then

$$T(au) = aT(u) = a0 = 0,$$

and so $au \in \text{null}\,(T)$.   □

Definition 6.2.5. The linear map $T : V \to W$ is called **injective** if, for all $u, v \in V$, the condition $Tu = Tv$ implies that $u = v$. In other words, different vectors in V are mapped to different vectors in W.

Proposition 6.2.6. *Let $T : V \to W$ be a linear map. Then T is injective if and only if* $\text{null}\,(T) = \{0\}$.

Proof.
("$\Longrightarrow$") Suppose that T is injective. Since $\text{null}\,(T)$ is a subspace of V, we know that $0 \in \text{null}\,(T)$. Assume that there is another vector $v \in V$ that is in the kernel. Then $T(v) = 0 = T(0)$. Since T is injective, this implies that $v = 0$, proving that $\text{null}\,(T) = \{0\}$.
("$\Longleftarrow$") Assume that $\text{null}\,(T) = \{0\}$, and let $u, v \in V$ be such that $Tu = Tv$. Then $0 = Tu - Tv = T(u - v)$ so that $u - v \in \text{null}\,(T)$. Hence $u - v = 0$, or, equivalently, $u = v$. This shows that T is indeed injective.   □

Example 6.2.7.

(1) The differentiation map $p(z) \mapsto p'(z)$ is not injective since $p'(z) = q'(z)$ implies that $p(z) = q(z) + c$, where $c \in \mathbb{F}$ is a constant.
(2) The identity map $I : V \to V$ is injective.
(3) The linear map $T : \mathbb{F}[z] \to \mathbb{F}[z]$ given by $T(p(z)) = z^2 p(z)$ is injective since it is easy to verify that $\text{null}\,(T) = \{0\}$.
(4) The linear map $T(x, y) = (x - 2y, 3x + y)$ is injective since $\text{null}\,(T) = \{(0, 0)\}$, as we calculated in Example 6.2.3.

6.3 Range

Definition 6.3.1. Let $T : V \to W$ be a linear map. The **range** of T, denoted by range (T), is the subset of vectors in W that are in the image of T. I.e.,

$$\text{range}\,(T) = \{Tv \mid v \in V\} = \{w \in W \mid \text{ there exists } v \in V \text{ such that } Tv = w\}.$$

Example 6.3.2. The range of the differentiation map $T : \mathbb{F}[z] \to \mathbb{F}[z]$ is range $(T) = \mathbb{F}[z]$ since, for every polynomial $q \in \mathbb{F}[z]$, there is a $p \in \mathbb{F}[z]$ such that $p' = q$.

Example 6.3.3. The range of the linear map $T(x, y) = (x - 2y, 3x + y)$ is $\mathbb{R}^2$ since, for any $(z_1, z_2) \in \mathbb{R}^2$, we have $T(x, y) = (z_1, z_2)$ if $(x, y) = \frac{1}{7}(z_1 + 2z_2, -3z_1 + z_2)$.

Proposition 6.3.4. *Let $T : V \to W$ be a linear map. Then* range (T) *is a subspace of W.*

Proof. We need to show that $0 \in$ range (T) and that range (T) is closed under addition and scalar multiplication. We already showed that $T0 = 0$ so that $0 \in$ range (T).

For closure under addition, let $w_1, w_2 \in$ range (T). Then there exist $v_1, v_2 \in V$ such that $Tv_1 = w_1$ and $Tv_2 = w_2$. Hence

$$T(v_1 + v_2) = Tv_1 + Tv_2 = w_1 + w_2,$$

and so $w_1 + w_2 \in$ range (T).

For closure under scalar multiplication, let $w \in$ range (T) and $a \in \mathbb{F}$. Then there exists a $v \in V$ such that $Tv = w$. Thus

$$T(av) = aTv = aw,$$

and so $aw \in$ range (T). $\qquad\square$

Definition 6.3.5. A linear map $T : V \to W$ is called **surjective** if range $(T) = W$. A linear map $T : V \to W$ is called **bijective** if T is both injective and surjective.

Example 6.3.6.

(1) The differentiation map $T : \mathbb{F}[z] \to \mathbb{F}[z]$ is surjective since range $(T) = \mathbb{F}[z]$. However, if we restrict ourselves to polynomials of degree at most m, then the differentiation map $T : \mathbb{F}_m[z] \to \mathbb{F}_m[z]$ is not surjective since polynomials of degree m are not in the range of T.
(2) The identity map $I : V \to V$ is surjective.
(3) The linear map $T : \mathbb{F}[z] \to \mathbb{F}[z]$ given by $T(p(z)) = z^2 p(z)$ is not surjective since, for example, there are no linear polynomials in the range of T.
(4) The linear map $T(x, y) = (x - 2y, 3x + y)$ is surjective since range $(T) = \mathbb{R}^2$, as we calculated in Example 6.3.3.

6.4   Homomorphisms

It should be mentioned that linear maps between vector spaces are also called **vector space homomorphisms**. Instead of the notation $\mathcal{L}(V, W)$, one often sees the convention

$$\mathrm{Hom}_{\mathbb{F}}(V, W) = \{T : V \to W \mid T \text{ is linear}\}.$$

A homomorphism $T : V \to W$ is also often called

- **Monomorphism** iff T is injective;
- **Epimorphism** iff T is surjective;
- **Isomorphism** iff T is bijective;
- **Endomorphism** iff $V = W$;
- **Automorphism** iff $V = W$ and T is bijective.

6.5   The dimension formula

The next theorem is the key result of this chapter. It relates the dimension of the kernel and range of a linear map.

Theorem 6.5.1. *Let V be a finite-dimensional vector space and $T : V \to W$ be a linear map. Then* $\mathrm{range}\,(T)$ *is a finite-dimensional subspace of W and*

$$\dim(V) = \dim(\mathrm{null}\,(T)) + \dim(\mathrm{range}\,(T)). \tag{6.4}$$

Proof. Let V be a finite-dimensional vector space and $T \in \mathcal{L}(V, W)$. Since $\mathrm{null}\,(T)$ is a subspace of V, we know that $\mathrm{null}\,(T)$ has a basis $(u_1, \ldots, u_m)$. This implies that $\dim(\mathrm{null}\,(T)) = m$. By the Basis Extension Theorem, it follows that $(u_1, \ldots, u_m)$ can be extended to a basis of V, say $(u_1, \ldots, u_m, v_1, \ldots, v_n)$, so that $\dim(V) = m + n$.

The theorem will follow by showing that $(Tv_1, \ldots, Tv_n)$ is a basis of $\mathrm{range}\,(T)$ since this would imply that $\mathrm{range}\,(T)$ is finite-dimensional and $\dim(\mathrm{range}\,(T)) = n$, proving Equation (6.4).

Since $(u_1, \ldots, u_m, v_1, \ldots, v_n)$ spans V, every $v \in V$ can be written as a linear combination of these vectors; i.e.,

$$v = a_1 u_1 + \cdots + a_m u_m + b_1 v_1 + \cdots + b_n v_n,$$

where $a_i, b_j \in \mathbb{F}$. Applying T to v, we obtain

$$Tv = b_1 Tv_1 + \cdots + b_n Tv_n,$$

where the terms Tu_i disappeared since $u_i \in \mathrm{null}\,(T)$. This shows that $(Tv_1, \ldots, Tv_n)$ indeed spans $\mathrm{range}\,(T)$.

To show that $(Tv_1, \ldots, Tv_n)$ is a basis of $\mathrm{range}\,(T)$, it remains to show that this list is linearly independent. Assume that $c_1, \ldots, c_n \in \mathbb{F}$ are such that

$$c_1 Tv_1 + \cdots + c_n Tv_n = 0.$$

By linearity of T, this implies that

$$T(c_1 v_1 + \cdots + c_n v_n) = 0,$$

and so $c_1 v_1 + \cdots + c_n v_n \in \text{null}\,(T)$. Since $(u_1, \ldots, u_m)$ is a basis of null (T), there must exist scalars $d_1, \ldots, d_m \in \mathbb{F}$ such that

$$c_1 v_1 + \cdots + c_n v_n = d_1 u_1 + \cdots + d_m u_m.$$

However, by the linear independence of $(u_1, \ldots, u_m, v_1, \ldots, v_n)$, this implies that all coefficients $c_1 = \cdots = c_n = d_1 = \cdots = d_m = 0$. Thus, $(Tv_1, \ldots, Tv_n)$ is linearly independent, and this completes the proof. $\qquad\square$

Example 6.5.2. Recall that the linear map $T : \mathbb{R}^2 \to \mathbb{R}^2$ defined by $T(x, y) = (x - 2y, 3x + y)$ has null $(T) = \{0\}$ and range $(T) = \mathbb{R}^2$. It follows that

$$\dim(\mathbb{R}^2) = 2 = 0 + 2 = \dim(\text{null}\,(T)) + \dim(\text{range}\,(T)).$$

Corollary 6.5.3. *Let $T \in \mathcal{L}(V, W)$.*

(1) If $\dim(V) > \dim(W)$, then T is not injective.
(2) If $\dim(V) < \dim(W)$, then T is not surjective.

Proof. By Theorem 6.5.1, we have that

$$\dim(\text{null}\,(T)) = \dim(V) - \dim(\text{range}\,(T))$$
$$\geq \dim(V) - \dim(W) > 0.$$

Since T is injective if and only if $\dim(\text{null}\,(T)) = 0$, T cannot be injective.
  Similarly,

$$\dim(\text{range}\,(T)) = \dim(V) - \dim(\text{null}\,(T))$$
$$\leq \dim(V) < \dim(W),$$

and so range (T) cannot be equal to W. Hence, T cannot be surjective. $\qquad\square$

6.6 The matrix of a linear map

Now we will see that every linear map $T \in \mathcal{L}(V, W)$, with V and W finite-dimensional vector spaces, can be encoded by a matrix, and, vice versa, every matrix defines such a linear map.

  Let V and W be finite-dimensional vector spaces, and let $T : V \to W$ be a linear map. Suppose that $(v_1, \ldots, v_n)$ is a basis of V and that $(w_1, \ldots, w_m)$ is a basis for W. We have seen in Theorem 6.1.3 that T is uniquely determined by specifying the vectors $Tv_1, \ldots, Tv_n \in W$. Since $(w_1, \ldots, w_m)$ is a basis of W, there exist unique scalars $a_{ij} \in \mathbb{F}$ such that

$$Tv_j = a_{1j} w_1 + \cdots + a_{mj} w_m \quad \text{for } 1 \leq j \leq n. \tag{6.5}$$

We can arrange these scalars in an $m \times n$ matrix as follows:

$$M(T) = \begin{bmatrix} a_{11} & \cdots & a_{1n} \\ \vdots & & \vdots \\ a_{m1} & \cdots & a_{mn} \end{bmatrix}.$$

Often, this is also written as $A = (a_{ij})_{1 \leq i \leq m, 1 \leq j \leq n}$. As in Section A.1.1, the set of all $m \times n$ matrices with entries in $\mathbb{F}$ is denoted by $\mathbb{F}^{m \times n}$.

Remark 6.6.1. It is important to remember that $M(T)$ not only depends on the linear map T but also on the choice of the bases $(v_1, \ldots, v_n)$ and $(w_1, \ldots, w_m)$ for V and W, respectively. The jth column of $M(T)$ contains the coefficients of the jth basis vector v_j when expanded in terms of the basis $(w_1, \ldots, w_m)$, as in Equation (6.5).

Example 6.6.2. Let $T : \mathbb{R}^2 \to \mathbb{R}^2$ be the linear map given by $T(x, y) = (ax + by, cx + dy)$ for some $a, b, c, d \in \mathbb{R}$. Then, with respect to the canonical basis of $\mathbb{R}^2$ given by $((1, 0), (0, 1))$, the corresponding matrix is

$$M(T) = \begin{bmatrix} a & b \\ c & d \end{bmatrix}$$

since $T(1, 0) = (a, c)$ gives the first column and $T(0, 1) = (b, d)$ gives the second column.

More generally, suppose that $V = \mathbb{F}^n$ and $W = \mathbb{F}^m$, and denote the standard basis for V by $(e_1, \ldots, e_n)$ and the standard basis for W by $(f_1, \ldots, f_m)$. Here, e_i (resp. f_i) is the n-tuple (resp. m-tuple) with a 1 in position i and zeroes everywhere else. Then the matrix $M(T) = (a_{ij})$ is given by

$$a_{ij} = (Te_j)_i,$$

where $(Te_j)_i$ denotes the ith component of the vector Te_j.

Example 6.6.3. Let $T : \mathbb{R}^2 \to \mathbb{R}^3$ be the linear map defined by $T(x, y) = (y, x + 2y, x + y)$. Then, with respect to the standard basis, we have $T(1, 0) = (0, 1, 1)$ and $T(0, 1) = (1, 2, 1)$ so that

$$M(T) = \begin{bmatrix} 0 & 1 \\ 1 & 2 \\ 1 & 1 \end{bmatrix}.$$

However, if alternatively we take the bases $((1, 2), (0, 1))$ for $\mathbb{R}^2$ and $((1, 0, 0), (0, 1, 0), (0, 0, 1))$ for $\mathbb{R}^3$, then $T(1, 2) = (2, 5, 3)$ and $T(0, 1) = (1, 2, 1)$ so that

$$M(T) = \begin{bmatrix} 2 & 1 \\ 5 & 2 \\ 3 & 1 \end{bmatrix}.$$

Example 6.6.4. Let $S : \mathbb{R}^2 \to \mathbb{R}^2$ be the linear map $S(x,y) = (y,x)$. With respect to the basis $((1,2),(0,1))$ for $\mathbb{R}^2$, we have

$$S(1,2) = (2,1) = 2(1,2) - 3(0,1) \quad \text{and} \quad S(0,1) = (1,0) = 1(1,2) - 2(0,1),$$

and so

$$M(S) = \begin{bmatrix} 2 & 1 \\ -3 & -2 \end{bmatrix}.$$

Given vector spaces V and W of dimensions n and m, respectively, and given a fixed choice of bases, note that there is a one-to-one correspondence between linear maps in $\mathcal{L}(V,W)$ and matrices in $\mathbb{F}^{m \times n}$. If we start with the linear map T, then the matrix $M(T) = A = (a_{ij})$ is defined via Equation (6.5). Conversely, given the matrix $A = (a_{ij}) \in \mathbb{F}^{m \times n}$, we can define a linear map $T : V \to W$ by setting

$$Tv_j = \sum_{i=1}^{m} a_{ij} w_i.$$

Recall that the set of linear maps $\mathcal{L}(V,W)$ is a vector space. Since we have a one-to-one correspondence between linear maps and matrices, we can also make the set of matrices $\mathbb{F}^{m \times n}$ into a vector space. Given two matrices $A = (a_{ij})$ and $B = (b_{ij})$ in $\mathbb{F}^{m \times n}$ and given a scalar $\alpha \in \mathbb{F}$, we define the **matrix addition** and **scalar multiplication** component-wise:

$$A + B = (a_{ij} + b_{ij}),$$
$$\alpha A = (\alpha a_{ij}).$$

Next, we show that the **composition** of linear maps imposes a product on matrices, also called **matrix multiplication**. Suppose U, V, W are vector spaces over $\mathbb{F}$ with bases $(u_1, \ldots, u_p)$, $(v_1, \ldots, v_n)$ and $(w_1, \ldots, w_m)$, respectively. Let $S : U \to V$ and $T : V \to W$ be linear maps. Then the product is a linear map $T \circ S : U \to W$.

Each linear map has its corresponding matrix $M(T) = A, M(S) = B$ and $M(TS) = C$. The question is whether C is determined by A and B. We have, for each $j \in \{1, 2, \ldots p\}$, that

$$(T \circ S)u_j = T(b_{1j}v_1 + \cdots + b_{nj}v_n) = b_{1j}Tv_1 + \cdots + b_{nj}Tv_n$$

$$= \sum_{k=1}^{n} b_{kj}Tv_k = \sum_{k=1}^{n} b_{kj}\left(\sum_{i=1}^{m} a_{ik}w_i\right)$$

$$= \sum_{i=1}^{m}\left(\sum_{k=1}^{n} a_{ik}b_{kj}\right)w_i.$$

Hence, the matrix $C = (c_{ij})$ is given by

$$c_{ij} = \sum_{k=1}^{n} a_{ik}b_{kj}. \tag{6.6}$$

Equation (6.6) can be used to define the $m \times p$ matrix C as the product of an $m \times n$ matrix A and an $n \times p$ matrix B, i.e.,

$$C = AB. \tag{6.7}$$

Our derivation implies that the correspondence between linear maps and matrices respects the product structure.

Proposition 6.6.5. *Let $S : U \to V$ and $T : V \to W$ be linear maps. Then*

$$M(TS) = M(T)M(S).$$

Example 6.6.6. With notation as in Examples 6.6.3 and 6.6.4, you should be able to verify that

$$M(TS) = M(T)M(S) = \begin{bmatrix} 2 & 1 \\ 5 & 2 \\ 3 & 1 \end{bmatrix} \begin{bmatrix} 2 & 1 \\ -3 & -2 \end{bmatrix} = \begin{bmatrix} 1 & 0 \\ 4 & 1 \\ 3 & 1 \end{bmatrix}.$$

Given a vector $v \in V$, we can also associate a matrix $M(v)$ to v as follows. Let $(v_1, \ldots, v_n)$ be a basis of V. Then there are unique scalars $b_1, \ldots, b_n$ such that

$$v = b_1 v_1 + \cdots b_n v_n.$$

The matrix of v is then defined to be the $n \times 1$ matrix

$$M(v) = \begin{bmatrix} b_1 \\ \vdots \\ b_n \end{bmatrix}.$$

Example 6.6.7. The matrix of a vector $x = (x_1, \ldots, x_n) \in \mathbb{F}^n$ in the standard basis $(e_1, \ldots, e_n)$ is the column vector or $n \times 1$ matrix

$$M(x) = \begin{bmatrix} x_1 \\ \vdots \\ x_n \end{bmatrix}$$

since $x = (x_1, \ldots, x_n) = x_1 e_1 + \cdots + x_n e_n$.

The next result shows how the notion of a matrix of a linear map $T : V \to W$ and the matrix of a vector $v \in V$ fit together.

Proposition 6.6.8. *Let $T : V \to W$ be a linear map. Then, for every $v \in V$,*

$$M(Tv) = M(T)M(v).$$

Proof. Let $(v_1, \ldots, v_n)$ be a basis of V and $(w_1, \ldots, w_m)$ be a basis for W. Suppose that, with respect to these bases, the matrix of T is $M(T) = (a_{ij})_{1 \leq i \leq m, 1 \leq j \leq n}$. This means that, for all $j \in \{1, 2, \ldots, n\}$,

$$Tv_j = \sum_{k=1}^{m} a_{kj} w_k.$$

The vector $v \in V$ can be written uniquely as a linear combination of the basis vectors as

$$v = b_1 v_1 + \cdots + b_n v_n.$$

Hence,

$$Tv = b_1 T v_1 + \cdots + b_n T v_n$$

$$= b_1 \sum_{k=1}^{m} a_{k1} w_k + \cdots + b_n \sum_{k=1}^{m} a_{kn} w_k$$

$$= \sum_{k=1}^{m} (a_{k1} b_1 + \cdots + a_{kn} b_n) w_k.$$

This shows that $M(Tv)$ is the $m \times 1$ matrix

$$M(Tv) = \begin{bmatrix} a_{11} b_1 + \cdots + a_{1n} b_n \\ \vdots \\ a_{m1} b_1 + \cdots + a_{mn} b_n \end{bmatrix}.$$

It is not hard to check, using the formula for matrix multiplication, that $M(T)M(v)$ gives the same result. $\square$

Example 6.6.9. Take the linear map S from Example 6.6.4 with basis $((1,2),(0,1))$ of $\mathbb{R}^2$. To determine the action on the vector $v = (1,4) \in \mathbb{R}^2$, note that $v = (1,4) = 1(1,2) + 2(0,1)$. Hence,

$$M(Sv) = M(S)M(v) = \begin{bmatrix} 2 & 1 \\ -3 & -2 \end{bmatrix} \begin{bmatrix} 1 \\ 2 \end{bmatrix} = \begin{bmatrix} 4 \\ -7 \end{bmatrix}.$$

This means that

$$Sv = 4(1,2) - 7(0,1) = (4,1),$$

which is indeed true.

6.7 Invertibility

Definition 6.7.1. A map $T : V \to W$ is called **invertible** if there exists a map $S : W \to V$ such that

$$TS = I_W \quad \text{and} \quad ST = I_V,$$

where $I_V : V \to V$ is the identity map on V and $I_W : W \to W$ is the identity map on W. We say that S is an **inverse** of T.

Note that if the map T is invertible, then the inverse is unique. Suppose S and R are inverses of T. Then

$$ST = I_V = RT,$$
$$TS = I_W = TR.$$

Hence,

$$S = S(TR) = (ST)R = R.$$

We denote the unique inverse of an invertible map T by T^{-1}.

Proposition 6.7.2. *A map* $T : V \longrightarrow W$ *is invertible if and only if* T *is injective and surjective.*

Proof.
("$\Longrightarrow$") Suppose T is invertible.

To show that T is injective, suppose that $u, v \in V$ are such that $Tu = Tv$. Apply the inverse T^{-1} of T to obtain $T^{-1}Tu = T^{-1}Tv$ so that $u = v$. Hence T is injective.

To show that T is surjective, we need to show that, for every $w \in W$, there is a $v \in V$ such that $Tv = w$. Take $v = T^{-1}w \in V$. Then $T(T^{-1}w) = w$. Hence T is surjective.

("$\Longleftarrow$") Suppose that T is injective and surjective. We need to show that T is invertible. We define a map $S \in \mathcal{L}(W, V)$ as follows. Since T is surjective, we know that, for every $w \in W$, there exists a $v \in V$ such that $Tv = w$. Moreover, since T is injective, this v is uniquely determined. Hence, define $Sw = v$.

We claim that S is the inverse of T. Note that, for all $w \in W$, we have $TSw = Tv = w$ so that $TS = I_W$. Similarly, for all $v \in V$, we have $STv = Sw = v$ so that $ST = I_V$. $\qquad\square$

Now we specialize to invertible *linear* maps.

Proposition 6.7.3. *Let* $T \in \mathcal{L}(V, W)$ *be invertible. Then* $T^{-1} \in \mathcal{L}(W, V)$.

Proof. Certainly $T^{-1} : W \longrightarrow V$ so we only need to show that T^{-1} is a linear map. For all $w_1, w_2 \in W$, we have

$$T(T^{-1}w_1 + T^{-1}w_2) = T(T^{-1}w_1) + T(T^{-1}w_2) = w_1 + w_2,$$

and so $T^{-1}w_1 + T^{-1}w_2$ is the unique vector v in V such that $Tv = w_1 + w_2 = w$. Hence,

$$T^{-1}w_1 + T^{-1}w_2 = v = T^{-1}w = T^{-1}(w_1 + w_2).$$

The proof that $T^{-1}(aw) = aT^{-1}w$ is similar. For $w \in W$ and $a \in \mathbb{F}$, we have

$$T(aT^{-1}w) = aT(T^{-1}w) = aw$$

so that $aT^{-1}w$ is the unique vector in V that maps to aw. Hence, $T^{-1}(aw) = aT^{-1}w$. $\qquad\square$

Example 6.7.4. The linear map $T(x, y) = (x - 2y, 3x + y)$ is both injective, since null $(T) = \{0\}$, and surjective, since range $(T) = \mathbb{R}^2$. Hence, T is invertible by Proposition 6.7.2.

Definition 6.7.5. Two vector spaces V and W are called **isomorphic** if there exists an invertible linear map $T \in \mathcal{L}(V, W)$.

Theorem 6.7.6. *Two finite-dimensional vector spaces* V *and* W *over* $\mathbb{F}$ *are isomorphic if and only if* $\dim(V) = \dim(W)$.

Proof.

("$\implies$") Suppose V and W are isomorphic. Then there exists an invertible linear map $T \in \mathcal{L}(V, W)$. Since T is invertible, it is injective and surjective, and so $\text{null}(T) = \{0\}$ and $\text{range}(T) = W$. Using the Dimension Formula, this implies that

$$\dim(V) = \dim(\text{null}(T)) + \dim(\text{range}(T)) = \dim(W).$$

("$\impliedby$") Suppose that $\dim(V) = \dim(W)$. Let $(v_1, \ldots, v_n)$ be a basis of V and $(w_1, \ldots, w_n)$ be a basis of W. Define the linear map $T : V \to W$ as

$$T(a_1 v_1 + \cdots + a_n v_n) = a_1 w_1 + \cdots + a_n w_n.$$

Since the scalars $a_1, \ldots, a_n \in \mathbb{F}$ are arbitrary and $(w_1, \ldots, w_n)$ spans W, this means that $\text{range}(T) = W$ and T is surjective. Also, since $(w_1, \ldots, w_n)$ is linearly independent, T is injective (since $a_1 w_1 + \cdots + a_n w_n = 0$ implies that all $a_1 = \cdots = a_n = 0$ and hence only the zero vector is mapped to zero). It follows that T is both injective and surjective; hence, by Proposition 6.7.2, T is invertible. Therefore, V and W are isomorphic. $\qquad\square$

We close this chapter by considering the case of linear maps having equal domain and codomain. As in Definition 6.1.1, a linear map $T \in \mathcal{L}(V, V)$ is called a **linear operator** on V. As the following remarkable theorem shows, the notions of injectivity, surjectivity, and invertibility of a linear operator T are the same — as long as V is finite-dimensional. A similar result does not hold for infinite-dimensional vector spaces. For example, the set of all polynomials $\mathbb{F}[z]$ is an infinite-dimensional vector space, and we saw that the differentiation map on $\mathbb{F}[z]$ is surjective but not injective.

Theorem 6.7.7. *Let V be a finite-dimensional vector space and $T : V \to V$ be a linear map. Then the following are equivalent:*

(1) T is invertible.
(2) T is injective.
(3) T is surjective.

Proof. By Proposition 6.7.2, Part 1 implies Part 2.

Next we show that Part 2 implies Part 3. If T is injective, then we know that $\text{null}(T) = \{0\}$. Hence, by the Dimension Formula, we have

$$\dim(\text{range}(T)) = \dim(V) - \dim(\text{null}(T)) = \dim(V).$$

Since $\text{range}(T) \subset V$ is a subspace of V, this implies that $\text{range}(T) = V$, and so T is surjective.

Finally, we show that Part 3 implies Part 1. Since T is surjective by assumption, we have $\text{range}(T) = V$. Thus, again by using the Dimension Formula,

$$\dim(\text{null}(T)) = \dim(V) - \dim(\text{range}(T)) = 0,$$

and so $\text{null}(T) = \{0\}$, from which T is injective. By Proposition 6.7.2, an injective and surjective linear map is invertible. $\qquad\square$

Exercises for Chapter 6

Calculational Exercises

(1) Define the map $T : \mathbb{R}^2 \to \mathbb{R}^2$ by $T(x, y) = (x + y, x)$.
   (a) Show that T is linear.
   (b) Show that T is surjective.
   (c) Find $\dim (\text{null} (T))$.
   (d) Find the matrix for T with respect to the canonical basis of $\mathbb{R}^2$.
   (e) Find the matrix for T with respect to the canonical basis for the domain $\mathbb{R}^2$ and the basis $((1, 1), (1, -1))$ for the target space $\mathbb{R}^2$.
   (f) Show that the map $F : \mathbb{R}^2 \to \mathbb{R}^2$ given by $F(x, y) = (x + y, x + 1)$ is not linear.

(2) Let $T \in \mathcal{L}(\mathbb{R}^2)$ be defined by

$$T \begin{pmatrix} x \\ y \end{pmatrix} = \begin{pmatrix} y \\ -x \end{pmatrix}, \quad \text{for all } \begin{pmatrix} x \\ y \end{pmatrix} \in \mathbb{R}^2 \, .$$

   (a) Show that T is surjective.
   (b) Find $\dim (\text{null} (T))$.
   (c) Find the matrix for T with respect to the canonical basis of $\mathbb{R}^2$.
   (d) Show that the map $F : \mathbb{R}^2 \to \mathbb{R}^2$ given by $F(x, y) = (x + y, x + 1)$ is not linear.

(3) Consider the complex vector spaces $\mathbb{C}^2$ and $\mathbb{C}^3$ with their canonical bases, and let $S \in \mathcal{L}(\mathbb{C}^3, \mathbb{C}^2)$ be the linear map defined by $S(v) = Av, \forall v \in \mathbb{C}^3$, where A is the matrix

$$A = M(S) = \begin{pmatrix} i & 1 & 1 \\ 2i & -1 & -1 \end{pmatrix} \, .$$

   Find a basis for null(S).

(4) Give an example of a function $f : \mathbb{R}^2 \to \mathbb{R}$ having the property that

$$\forall \, a \in \mathbb{R}, \forall \, v \in \mathbb{R}^2, f(av) = af(v)$$

   but such that f is not a linear map.

(5) Show that the linear map $T : \mathbb{F}^4 \to \mathbb{F}^2$ is surjective if

$$\text{null}(T) = \{(x_1, x_2, x_3, x_4) \in \mathbb{F}^4 \mid x_1 = 5x_2, x_3 = 7x_4\}.$$

(6) Show that no linear map $T : \mathbb{F}^5 \to \mathbb{F}^2$ can have as its null space the set

$$\{(x_1, x_2, x_3, x_4, x_5) \in \mathbb{F}^5 \mid x_1 = 3x_2, x_3 = x_4 = x_5\}.$$

(7) Describe the set of solutions $x = (x_1, x_2, x_3) \in \mathbb{R}^3$ of the system of equations

$$\left. \begin{array}{r} x_1 - x_2 + x_3 = 0 \\ x_1 + 2x_2 + x_3 = 0 \\ 2x_1 + x_2 + 2x_3 = 0 \end{array} \right\} \, .$$

Proof-Writing Exercises

(1) Let V and W be vector spaces over $\mathbb{F}$ with V finite-dimensional, and let U be any subspace of V. Given a linear map $S \in \mathcal{L}(U, W)$, prove that there exists a linear map $T \in \mathcal{L}(V, W)$ such that, for every $u \in U$, $S(u) = T(u)$.

(2) Let V and W be vector spaces over $\mathbb{F}$, and suppose that $T \in \mathcal{L}(V, W)$ is injective. Given a linearly independent list $(v_1, \ldots, v_n)$ of vectors in V, prove that the list $(T(v_1), \ldots, T(v_n))$ is linearly independent in W.

(3) Let U, V, and W be vector spaces over $\mathbb{F}$, and suppose that the linear maps $S \in \mathcal{L}(U, V)$ and $T \in \mathcal{L}(V, W)$ are both injective. Prove that the composition map $T \circ S$ is injective.

(4) Let V and W be vector spaces over $\mathbb{F}$, and suppose that $T \in \mathcal{L}(V, W)$ is surjective. Given a spanning list $(v_1, \ldots, v_n)$ for V, prove that

$$\text{span}(T(v_1), \ldots, T(v_n)) = W.$$

(5) Let V and W be vector spaces over $\mathbb{F}$ with V finite-dimensional. Given $T \in \mathcal{L}(V, W)$, prove that there is a subspace U of V such that

$$U \cap \text{null}(T) = \{0\} \quad \text{and} \quad \text{range}(T) = \{T(u) \mid u \in U\}.$$

(6) Let V be a vector space over $\mathbb{F}$, and suppose that there is a linear map $T \in \mathcal{L}(V, V)$ such that both $\text{null}(T)$ and $\text{range}(T)$ are finite-dimensional subspaces of V. Prove that V must also be finite-dimensional.

(7) Let U, V, and W be finite-dimensional vector spaces over $\mathbb{F}$ with $S \in \mathcal{L}(U, V)$ and $T \in \mathcal{L}(V, W)$. Prove that

$$\dim(\text{null}(T \circ S)) \leq \dim(\text{null}(T)) + \dim(\text{null}(S)).$$

(8) Let V be a finite-dimensional vector space over $\mathbb{F}$ with $S, T \in \mathcal{L}(V, V)$. Prove that $T \circ S$ is invertible if and only if both S and T are invertible.

(9) Let V be a finite-dimensional vector space over $\mathbb{F}$ with $S, T \in \mathcal{L}(V, V)$, and denote by I the identity map on V. Prove that $T \circ S = I$ if and only if $S \circ T = I$.

Chapter 7

Eigenvalues and Eigenvectors

In this chapter we study linear operators $T : V \to V$ on a finite-dimensional vector space V. We are interested in finding bases B for V such that the matrix $M(T)$ of T with respect to B is upper triangular or, if possible, diagonal. This quest leads us to the notions of eigenvalues and eigenvectors of a linear operator, which is one of the most important concepts in Linear Algebra and essential for many of its applications. For example, quantum mechanics is largely based upon the study of eigenvalues and eigenvectors of operators on finite- and infinite-dimensional vector spaces.

7.1 Invariant subspaces

To begin our study, we will look at subspaces U of V that have special properties under an operator $T \in \mathcal{L}(V, V)$.

Definition 7.1.1. Let V be a finite-dimensional vector space over $\mathbb{F}$ with $\dim(V) \geq 1$, and let $T \in \mathcal{L}(V, V)$ be an operator in V. Then a subspace $U \subset V$ is called an **invariant subspace under** T if

$$Tu \in U \quad \text{for all } u \in U.$$

That is, U is invariant under T if the image of every vector in U under T remains within U. We denote this as $TU = \{Tu \mid u \in U\} \subset U$.

Example 7.1.2. The subspaces $\text{null}\,(T)$ and $\text{range}\,(T)$ are invariant subspaces under T. To see this, let $u \in \text{null}\,(T)$. This means that $Tu = 0$. But, since $0 \in \text{null}\,(T)$, this implies that $Tu = 0 \in \text{null}\,(T)$. Similarly, let $u \in \text{range}\,(T)$. Since $Tv \in \text{range}\,(T)$ for all $v \in V$, in particular we have $Tu \in \text{range}\,(T)$.

Example 7.1.3. Take the linear operator $T : \mathbb{R}^3 \to \mathbb{R}^3$ corresponding to the matrix

$$\begin{bmatrix} 1 & 2 & 0 \\ 1 & 1 & 0 \\ 0 & 0 & 2 \end{bmatrix}$$

with respect to the basis (e_1, e_2, e_3). Then $\text{span}(e_1, e_2)$ and $\text{span}(e_3)$ are both invariant subspaces under T.

An important special case of Definition 7.1.1 is that of one-dimensional invariant subspaces under an operator $T \in \mathcal{L}(V, V)$. If $\dim(U) = 1$, then there exists a non-zero vector $u \in V$ such that

$$U = \{au \mid a \in \mathbb{F}\}.$$

In this case, we must have

$$Tu = \lambda u \quad \text{for some } \lambda \in \mathbb{F}.$$

This motivates the definitions of eigenvectors and eigenvalues of a linear operator, as given in the next section.

7.2 Eigenvalues

Definition 7.2.1. Let $T \in \mathcal{L}(V, V)$. Then $\lambda \in \mathbb{F}$ is an **eigenvalue** of T if there exists a non-zero vector $u \in V$ such that

$$Tu = \lambda u.$$

The vector u is called an **eigenvector** of T **corresponding to** the eigenvalue λ.

Finding the eigenvalues and eigenvectors of a linear operator is one of the most important problems in Linear Algebra. We will see later that this so-called "eigen-information" has many uses and applications. (As an example, quantum mechanics is based upon understanding the eigenvalues and eigenvectors of operators on specifically defined vector spaces. These vector spaces are often infinite-dimensional, though, and so we do not consider them further in this book.)

Example 7.2.2.

(1) Let T be the zero map defined by $T(v) = 0$ for all $v \in V$. Then every vector $u \neq 0$ is an eigenvector of T with eigenvalue 0.
(2) Let I be the identity map defined by $I(v) = v$ for all $v \in V$. Then every vector $u \neq 0$ is an eigenvector of T with eigenvalue 1.
(3) The projection map $P : \mathbb{R}^3 \to \mathbb{R}^3$ defined by $P(x, y, z) = (x, y, 0)$ has eigenvalues 0 and 1. The vector $(0, 0, 1)$ is an eigenvector with eigenvalue 0, and both $(1, 0, 0)$ and $(0, 1, 0)$ are eigenvectors with eigenvalue 1.
(4) Take the operator $R : \mathbb{F}^2 \to \mathbb{F}^2$ defined by $R(x, y) = (-y, x)$. When $\mathbb{F} = \mathbb{R}$, R can be interpreted as counterclockwise rotation by $90°$. From this interpretation, it is clear that no non-zero vector in $\mathbb{R}^2$ is mapped to a scalar multiple of itself. Hence, for $\mathbb{F} = \mathbb{R}$, the operator R has no eigenvalues.
For $\mathbb{F} = \mathbb{C}$, though, the situation is significantly different! In this case, $\lambda \in \mathbb{C}$ is an eigenvalue of R if

$$R(x, y) = (-y, x) = \lambda(x, y)$$

so that $y = -\lambda x$ and $x = \lambda y$. This implies that $y = -\lambda^2 y$, i.e., $\lambda^2 = -1$. The solutions are hence $\lambda = \pm i$. One can check that $(1, -i)$ is an eigenvector with eigenvalue i and that $(1, i)$ is an eigenvector with eigenvalue $-i$.

Eigenspaces are important examples of invariant subspaces. Let $T \in \mathcal{L}(V, V)$, and let $\lambda \in \mathbb{F}$ be an eigenvalue of T. Then

$$V_\lambda = \{v \in V \mid Tv = \lambda v\}$$

is called an **eigenspace** of T. Equivalently,

$$V_\lambda = \text{null}\,(T - \lambda I).$$

Note that $V_\lambda \neq \{0\}$ since λ is an eigenvalue if and only if there exists a non-zero vector $u \in V$ such that $Tu = \lambda u$. We can reformulate this as follows:

- $\lambda \in \mathbb{F}$ is an eigenvalue of T if and only if the operator $T - \lambda I$ is not injective.

Since the notion of injectivity, surjectivity, and invertibility are equivalent for operators on a finite-dimensional vector space, we can equivalently say either of the following:

- $\lambda \in \mathbb{F}$ is an eigenvalue of T if and only if the operator $T - \lambda I$ is not surjective.
- $\lambda \in \mathbb{F}$ is an eigenvalue of T if and only if the operator $T - \lambda I$ is not invertible.

We close this section with two fundamental facts about eigenvalues and eigenvectors.

Theorem 7.2.3. *Let $T \in \mathcal{L}(V, V)$, and let $\lambda_1, \ldots, \lambda_m \in \mathbb{F}$ be m distinct eigenvalues of T with corresponding non-zero eigenvectors $v_1, \ldots, v_m$. Then $(v_1, \ldots, v_m)$ is linearly independent.*

Proof. Suppose that $(v_1, \ldots, v_m)$ is linearly dependent. Then, by the Linear Dependence Lemma, there exists an index $k \in \{2, \ldots, m\}$ such that

$$v_k \in \text{span}(v_1, \ldots, v_{k-1})$$

and such that $(v_1, \ldots, v_{k-1})$ is linearly independent. This means that there exist scalars $a_1, \ldots, a_{k-1} \in \mathbb{F}$ such that

$$v_k = a_1 v_1 + \cdots + a_{k-1} v_{k-1}. \tag{7.1}$$

Applying T to both sides yields, using the fact that v_j is an eigenvector with eigenvalue λ_j,

$$\lambda_k v_k = a_1 \lambda_1 v_1 + \cdots + a_{k-1} \lambda_{k-1} v_{k-1}.$$

Subtracting λ_k times Equation (7.1) from this, we obtain

$$0 = (\lambda_k - \lambda_1) a_1 v_1 + \cdots + (\lambda_k - \lambda_{k-1}) a_{k-1} v_{k-1}.$$

Since $(v_1, \ldots, v_{k-1})$ is linearly independent, we must have $(\lambda_k - \lambda_j) a_j = 0$ for all $j = 1, 2, \ldots, k - 1$. By assumption, all eigenvalues are distinct, so $\lambda_k - \lambda_j \neq 0$, which implies that $a_j = 0$ for all $j = 1, 2, \ldots, k - 1$. But then, by Equation (7.1), $v_k = 0$, which contradicts the assumption that all eigenvectors are non-zero. Hence $(v_1, \ldots, v_m)$ is linearly independent. $\square$

Corollary 7.2.4. *Any operator $T \in \mathcal{L}(V, V)$ has at most* $\dim(V)$ *distinct eigenvalues.*

Proof. Let $\lambda_1, \ldots, \lambda_m$ be distinct eigenvalues of T, and let $v_1, \ldots, v_m$ be corresponding non-zero eigenvectors. By Theorem 7.2.3, the list $(v_1, \ldots, v_m)$ is linearly independent. Hence $m \leq \dim(V)$.   □

7.3   Diagonal matrices

Note that if T has $n = \dim(V)$ distinct eigenvalues, then there exists a basis $(v_1, \ldots, v_n)$ of V such that

$$Tv_j = \lambda_j v_j, \quad \text{for all } j = 1, 2, \ldots, n.$$

Then any $v \in V$ can be written as a linear combination $v = a_1 v_1 + \cdots + a_n v_n$ of $v_1, \ldots, v_n$. Applying T to this, we obtain

$$Tv = \lambda_1 a_1 v_1 + \cdots + \lambda_n a_n v_n.$$

Hence the vector

$$M(v) = \begin{bmatrix} a_1 \\ \vdots \\ a_n \end{bmatrix}$$

is mapped to

$$M(Tv) = \begin{bmatrix} \lambda_1 a_1 \\ \vdots \\ \lambda_n a_n \end{bmatrix}.$$

This means that the matrix $M(T)$ for T with respect to the basis of eigenvectors $(v_1, \ldots, v_n)$ is **diagonal**, and so we call T **diagonalizable**:

$$M(T) = \begin{bmatrix} \lambda_1 & & 0 \\ & \ddots & \\ 0 & & \lambda_n \end{bmatrix}.$$

We summarize the results of the above discussion in the following Proposition.

Proposition 7.3.1. *If $T \in \mathcal{L}(V, V)$ has* $\dim(V)$ *distinct eigenvalues, then $M(T)$ is diagonal with respect to some basis of V. Moreover, V has a basis consisting of eigenvectors of T.*

7.4 Existence of eigenvalues

In what follows, we want to study the question of when eigenvalues exist for a given operator T. To answer this question, we will use polynomials $p(z) \in \mathbb{F}[z]$ evaluated on operators $T \in \mathcal{L}(V, V)$ (or, equivalently, on square matrices $A \in \mathbb{F}^{n \times n}$). More explicitly, given a polynomial

$$p(z) = a_0 + a_1 z + \cdots + a_k z^k$$

we can associate the operator

$$p(T) = a_0 I_V + a_1 T + \cdots + a_k T^k.$$

Note that, for $p(z), q(z) \in \mathbb{F}[z]$, we have

$$(pq)(T) = p(T)q(T) = q(T)p(T).$$

The results of this section will be for complex vector spaces. This is because the proof of the existence of eigenvalues relies on the Fundamental Theorem of Algebra from Chapter 3, which makes a statement about the existence of zeroes of polynomials over $\mathbb{C}$.

Theorem 7.4.1. *Let $V \neq \{0\}$ be a finite-dimensional vector space over $\mathbb{C}$, and let $T \in \mathcal{L}(V, V)$. Then T has at least one eigenvalue.*

Proof. Let $v \in V$ with $v \neq 0$, and consider the list of vectors

$$(v, Tv, T^2 v, \ldots, T^n v),$$

where $n = \dim(V)$. Since the list contains $n + 1$ vectors, it must be linearly dependent. Hence, there exist scalars $a_0, a_1, \ldots, a_n \in \mathbb{C}$, not all zero, such that

$$0 = a_0 v + a_1 T v + a_2 T^2 v + \cdots + a_n T^n v.$$

Let m be the largest index for which $a_m \neq 0$. Since $v \neq 0$, we must have $m > 0$ (but possibly $m = n$). Consider the polynomial

$$p(z) = a_0 + a_1 z + \cdots + a_m z^m.$$

By Theorem 3.2.2 (3) it can be factored as

$$p(z) = c(z - \lambda_1) \cdots (z - \lambda_m),$$

where $c, \lambda_1, \ldots, \lambda_m \in \mathbb{C}$ and $c \neq 0$.

Therefore,

$$0 = a_0 v + a_1 T v + a_2 T^2 v + \cdots + a_n T^n v = p(T)v$$
$$= c(T - \lambda_1 I)(T - \lambda_2 I) \cdots (T - \lambda_m I)v,$$

and so at least one of the factors $T - \lambda_j I$ must be non-injective. In other words, this λ_j is an eigenvalue of T. $\square$

Note that the proof of Theorem 7.4.1 only uses basic concepts about linear maps, which is the same approach as in a popular textbook called *Linear Algebra Done Right* by Sheldon Axler. Many other textbooks rely on significantly more difficult proofs using concepts like the determinant and characteristic polynomial of a matrix. At the same time, it is often preferable to use the characteristic polynomial of a matrix in order to compute eigen-information of an operator; we discuss this approach in Chapter 8.

Note also that Theorem 7.4.1 does not hold for real vector spaces. E.g., as we saw in Example 7.2.2, the rotation operator R on $\mathbb{R}^2$ has no eigenvalues.

7.5   Upper triangular matrices

As before, let V be a complex vector space.

Let $T \in \mathcal{L}(V, V)$ and $(v_1, \ldots, v_n)$ be a basis for V. Recall that we can associate a matrix $M(T) \in \mathbb{C}^{n \times n}$ to the operator T. By Theorem 7.4.1, we know that T has at least one eigenvalue, say $\lambda \in \mathbb{C}$. Let $v_1 \neq 0$ be an eigenvector corresponding to λ. By the Basis Extension Theorem, we can extend the list (v_1) to a basis of V. Since $Tv_1 = \lambda v_1$, the first column of $M(T)$ with respect to this basis is

$$\begin{bmatrix} \lambda \\ 0 \\ \vdots \\ 0 \end{bmatrix}.$$

What we will show next is that we can find a basis of V such that the matrix $M(T)$ is **upper triangular**.

Definition 7.5.1. A matrix $A = (a_{ij}) \in \mathbb{F}^{n \times n}$ is called **upper triangular** if $a_{ij} = 0$ for $i > j$.

Schematically, an upper triangular matrix has the form

$$\begin{bmatrix} * & & * \\ & \ddots & \\ 0 & & * \end{bmatrix},$$

where the entries $*$ can be anything and every entry below the main diagonal is zero.

Here are two reasons why having an operator T represented by an upper triangular matrix can be quite convenient:

(1) the eigenvalues are on the diagonal (as we will see later);
(2) it is easy to solve the corresponding system of linear equations by back substitution (as discussed in Section A.3).

The next proposition tells us what upper triangularity means in terms of linear operators and invariant subspaces.

Proposition 7.5.2. *Suppose $T \in \mathcal{L}(V, V)$ and that $(v_1, \ldots, v_n)$ is a basis of V. Then the following statements are equivalent:*

(1) the matrix $M(T)$ with respect to the basis $(v_1, \ldots, v_n)$ is upper triangular;
(2) $Tv_k \in \text{span}(v_1, \ldots, v_k)$ for each $k = 1, 2, \ldots, n$;
(3) $\text{span}(v_1, \ldots, v_k)$ is invariant under T for each $k = 1, 2, \ldots, n$.

Proof. The equivalence of Condition 1 and Condition 2 follows easily from the definition since Condition 2 implies that the matrix elements below the diagonal are zero.

Clearly, Condition 3 implies Condition 2. To show that Condition 2 implies Condition 3, note that any vector $v \in \text{span}(v_1, \ldots, v_k)$ can be written as $v = a_1 v_1 + \cdots + a_k v_k$. Applying T, we obtain

$$Tv = a_1 Tv_1 + \cdots + a_k Tv_k \in \text{span}(v_1, \ldots, v_k)$$

since, by Condition 2, each $Tv_j \in \text{span}(v_1, \ldots, v_j) \subset \text{span}(v_1, \ldots, v_k)$ for $j = 1, 2, \ldots, k$ and since the span is a subspace of V. $\qquad\square$

The next theorem shows that complex vector spaces indeed have some basis for which the matrix of a given operator is upper triangular.

Theorem 7.5.3. *Let V be a finite-dimensional vector space over $\mathbb{C}$ and $T \in \mathcal{L}(V, V)$. Then there exists a basis B for V such that $M(T)$ is upper triangular with respect to B.*

Proof. We proceed by induction on $\dim(V)$. If $\dim(V) = 1$, then there is nothing to prove.

Hence, assume that $\dim(V) = n > 1$ and that we have proven the result of the theorem for all $T \in \mathcal{L}(W, W)$, where W is a complex vector space with $\dim(W) \leq n - 1$. By Theorem 7.4.1, T has at least one eigenvalue λ. Define

$$U = \text{range}\,(T - \lambda I),$$

and note that

(1) $\dim(U) < \dim(V) = n$ since λ is an eigenvalue of T and hence $T - \lambda I$ is not surjective;
(2) U is an invariant subspace of T since, for all $u \in U$, we have

$$Tu = (T - \lambda I)u + \lambda u,$$

which implies that $Tu \in U$ since $(T - \lambda I)u \in \text{range}\,(T - \lambda I) = U$ and $\lambda u \in U$.

Therefore, we may consider the operator $S = T|_U$, which is the operator obtained by restricting T to the subspace U. By the induction hypothesis, there exists a basis $(u_1, \ldots, u_m)$ of U with $m \leq n - 1$ such that $M(S)$ is upper triangular with respect to $(u_1, \ldots, u_m)$. This means that

$$Tu_j = Su_j \in \text{span}(u_1, \ldots, u_j), \quad \text{for all } j = 1, 2, \ldots, m.$$

Extend this to a basis $(u_1, \ldots, u_m, v_1, \ldots, v_k)$ of V. Then

$$Tv_j = (T - \lambda I)v_j + \lambda v_j, \quad \text{for all } j = 1, 2, \ldots, k.$$

Since $(T - \lambda I)v_j \in \text{range}\,(T - \lambda I) = U = \text{span}(u_1, \ldots, u_m)$, we have that

$$Tv_j \in \text{span}(u_1, \ldots, u_m, v_1, \ldots, v_j), \quad \text{for all } j = 1, 2, \ldots, k.$$

Hence, T is upper triangular with respect to the basis $(u_1, \ldots, u_m, v_1, \ldots, v_k)$. $\quad \square$

The following are two very important facts about upper triangular matrices and their associated operators.

Proposition 7.5.4. *Suppose $T \in \mathcal{L}(V, V)$ is a linear operator and that $M(T)$ is upper triangular with respect to some basis of V. Then*

(1) T is invertible if and only if all entries on the diagonal of $M(T)$ are non-zero.
(2) The eigenvalues of T are precisely the diagonal elements of $M(T)$.

Proof of Proposition 7.5.4, Part 1. Let $(v_1, \ldots, v_n)$ be a basis of V such that

$$M(T) = \begin{bmatrix} \lambda_1 & & * \\ & \ddots & \\ 0 & & \lambda_n \end{bmatrix}$$

is upper triangular. The claim is that T is invertible if and only if $\lambda_k \neq 0$ for all $k = 1, 2, \ldots, n$. Equivalently, this can be reformulated as follows: T is not invertible if and only if $\lambda_k = 0$ for at least one $k \in \{1, 2, \ldots, n\}$.

Suppose $\lambda_k = 0$. We will show that this implies the non-invertibility of T. If $k = 1$, this is obvious since then $Tv_1 = 0$, which implies that $v_1 \in \text{null}\,(T)$ so that T is not injective and hence not invertible. So assume that $k > 1$. Then

$$Tv_j \in \text{span}(v_1, \ldots, v_{k-1}), \quad \text{for all } j \leq k,$$

since T is upper triangular and $\lambda_k = 0$. Hence, we may define $S = T|_{\text{span}(v_1, \ldots, v_k)}$ to be the restriction of T to the subspace $\text{span}(v_1, \ldots, v_k)$ so that

$$S : \text{span}(v_1, \ldots, v_k) \to \text{span}(v_1, \ldots, v_{k-1}).$$

The linear map S is not injective since the dimension of the domain is larger than the dimension of its codomain, i.e.,

$$\dim(\text{span}(v_1, \ldots, v_k)) = k > k - 1 = \dim(\text{span}(v_1, \ldots, v_{k-1})).$$

Hence, there exists a vector $0 \neq v \in \text{span}(v_1, \ldots, v_k)$ such that $Sv = Tv = 0$. This implies that T is also not injective and therefore not invertible.

Now suppose that T is not invertible. We need to show that at least one $\lambda_k = 0$. The linear map T not being invertible implies that T is not injective. Hence, there exists a vector $0 \neq v \in V$ such that $Tv = 0$, and we can write

$$v = a_1 v_1 + \cdots + a_k v_k$$

for some k, where $a_k \neq 0$. Then

$$0 = Tv = (a_1 T v_1 + \cdots + a_{k-1} T v_{k-1}) + a_k T v_k. \tag{7.2}$$

Since T is upper triangular with respect to the basis $(v_1, \ldots, v_n)$, we know that $a_1 T v_1 + \cdots + a_{k-1} T v_{k-1} \in \text{span}(v_1, \ldots, v_{k-1})$. Hence, Equation (7.2) shows that $T v_k \in \text{span}(v_1, \ldots, v_{k-1})$, which implies that $\lambda_k = 0$. $\square$

Proof of Proposition 7.5.4, Part 2. Recall that $\lambda \in \mathbb{F}$ is an eigenvalue of T if and only if the operator $T - \lambda I$ is not invertible. Let $(v_1, \ldots, v_n)$ be a basis such that $M(T)$ is upper triangular. Then

$$M(T - \lambda I) = \begin{bmatrix} \lambda_1 - \lambda & & * \\ & \ddots & \\ 0 & & \lambda_n - \lambda \end{bmatrix}.$$

Hence, by Proposition 7.5.4(1), $T - \lambda I$ is not invertible if and only if $\lambda = \lambda_k$ for some k. $\square$

7.6 Diagonalization of 2×2 matrices and applications

Let $A = \begin{bmatrix} a & b \\ c & d \end{bmatrix} \in \mathbb{F}^{2 \times 2}$, and recall that we can define a linear operator $T \in \mathcal{L}(\mathbb{F}^2)$ on $\mathbb{F}^2$ by setting $T(v) = Av$ for each $v = \begin{bmatrix} v_1 \\ v_2 \end{bmatrix} \in \mathbb{F}^2$.

One method for finding the eigen-information of T is to analyze the solutions of the matrix equation $Av = \lambda v$ for $\lambda \in \mathbb{F}$ and $v \in \mathbb{F}^2$. In particular, using the definition of eigenvector and eigenvalue, v is an eigenvector associated to the eigenvalue λ if and only if $Av = T(v) = \lambda v$.

A simpler method involves the equivalent matrix equation $(A - \lambda I)v = 0$, where I denotes the identity map on $\mathbb{F}^2$. In particular, $0 \neq v \in \mathbb{F}^2$ is an eigenvector for T associated to the eigenvalue $\lambda \in \mathbb{F}$ if and only if the system of linear equations

$$\left. \begin{array}{r} (a - \lambda)v_1 + bv_2 = 0 \\ cv_1 + (d - \lambda)v_2 = 0 \end{array} \right\} \tag{7.3}$$

has a non-trivial solution. Moreover, System (7.3) has a non-trivial solution if and only if the polynomial $p(\lambda) = (a - \lambda)(d - \lambda) - bc$ evaluates to zero. (See Proof-writing Exercise 12 on page 79.)

In other words, the eigenvalues for T are exactly the $\lambda \in \mathbb{F}$ for which $p(\lambda) = 0$, and the eigenvectors for T associated to an eigenvalue λ are exactly the non-zero vectors $v = \begin{bmatrix} v_1 \\ v_2 \end{bmatrix} \in \mathbb{F}^2$ that satisfy System (7.3).

Example 7.6.1. Let $A = \begin{bmatrix} -2 & -1 \\ 5 & 2 \end{bmatrix}$. Then $p(\lambda) = (-2 - \lambda)(2 - \lambda) - (-1)(5) = \lambda^2 + 1$, which is equal to zero exactly when $\lambda = \pm i$. Moreover, if $\lambda = i$, then the System (7.3) becomes

$$\left. \begin{array}{rl} (-2 - i)v_1 - & v_2 = 0 \\ 5v_1 + (2 - i)v_2 = 0 \end{array} \right\},$$

which is satisfied by any vector $v = \begin{bmatrix} v_1 \\ v_2 \end{bmatrix} \in \mathbb{C}^2$ such that $v_2 = (-2 - i)v_1$. Similarly, if $\lambda = -i$, then the System (7.3) becomes

$$\left. \begin{array}{rl} (-2 + i)v_1 - & v_2 = 0 \\ 5v_1 + (2 + i)v_2 = 0 \end{array} \right\},$$

which is satisfied by any vector $v = \begin{bmatrix} v_1 \\ v_2 \end{bmatrix} \in \mathbb{C}^2$ such that $v_2 = (-2 + i)v_1$.

It follows that, given $A = \begin{bmatrix} -2 & -1 \\ 5 & 2 \end{bmatrix}$, the linear operator on $\mathbb{C}^2$ defined by $T(v) = Av$ has eigenvalues $\lambda = \pm i$, with associated eigenvectors as described above.

Example 7.6.2. Take the rotation $R_\theta : \mathbb{R}^2 \to \mathbb{R}^2$ by an angle $\theta \in [0, 2\pi)$ given by the matrix

$$R_\theta = \begin{bmatrix} \cos \theta & -\sin \theta \\ \sin \theta & \cos \theta \end{bmatrix}.$$

Then we obtain the eigenvalues by solving the polynomial equation

$$\begin{aligned} p(\lambda) &= (\cos \theta - \lambda)^2 + \sin^2 \theta \\ &= \lambda^2 - 2\lambda \cos \theta + 1 = 0, \end{aligned}$$

where we have used the fact that $\sin^2 \theta + \cos^2 \theta = 1$. Solving for λ in $\mathbb{C}$, we obtain

$$\lambda = \cos \theta \pm \sqrt{\cos^2 \theta - 1} = \cos \theta \pm \sqrt{-\sin^2 \theta} = \cos \theta \pm i \sin \theta = e^{\pm i\theta}.$$

We see that, as an operator over the real vector space $\mathbb{R}^2$, the operator R_θ only has eigenvalues when $\theta = 0$ or $\theta = \pi$. However, if we interpret the vector $\begin{bmatrix} x_1 \\ x_2 \end{bmatrix} \in \mathbb{R}^2$ as a complex number $z = x_1 + ix_2$, then z is an eigenvector if $R_\theta : \mathbb{C} \to \mathbb{C}$ maps $z \mapsto \lambda z = e^{\pm i\theta} z$. Moreover, from Section 2.3.2, we know that multiplication by $e^{\pm i\theta}$ corresponds to rotation by the angle $\pm \theta$.

Exercises for Chapter 7

Calculational Exercises

(1) Let $T \in \mathcal{L}(\mathbb{F}^2, \mathbb{F}^2)$ be defined by

$$T(u, v) = (v, u)$$

for every $u, v \in \mathbb{F}$. Compute the eigenvalues and associated eigenvectors for T.

(2) Let $T \in \mathcal{L}(\mathbb{F}^3, \mathbb{F}^3)$ be defined by

$$T(u, v, w) = (2v, 0, 5w)$$

for every $u, v, w \in \mathbb{F}$. Compute the eigenvalues and associated eigenvectors for T.

(3) Let $n \in \mathbb{Z}_+$ be a positive integer and $T \in \mathcal{L}(\mathbb{F}^n, \mathbb{F}^n)$ be defined by

$$T(x_1, \ldots, x_n) = (x_1 + \cdots + x_n, \ldots, x_1 + \cdots + x_n)$$

for every $x_1, \ldots, x_n \in \mathbb{F}$. Compute the eigenvalues and associated eigenvectors for T.

(4) Find eigenvalues and associated eigenvectors for the linear operators on $\mathbb{F}^2$ defined by each given 2×2 matrix.

(a) $\begin{bmatrix} 3 & 0 \\ 8 & -1 \end{bmatrix}$ (b) $\begin{bmatrix} 10 & -9 \\ 4 & -2 \end{bmatrix}$ (c) $\begin{bmatrix} 0 & 3 \\ 4 & 0 \end{bmatrix}$

(d) $\begin{bmatrix} -2 & -7 \\ 1 & 2 \end{bmatrix}$ (e) $\begin{bmatrix} 0 & 0 \\ 0 & 0 \end{bmatrix}$ (f) $\begin{bmatrix} 1 & 0 \\ 0 & 1 \end{bmatrix}$

Hint: Use the fact that, given a matrix $A = \begin{bmatrix} a & b \\ c & d \end{bmatrix} \in \mathbb{F}^{2\times 2}$, $\lambda \in \mathbb{F}$ is an eigenvalue for A if and only if $(a - \lambda)(d - \lambda) - bc = 0$.

(5) For each matrix A below, find eigenvalues for the induced linear operator T on $\mathbb{F}^n$ without performing any calculations. Then describe the eigenvectors $v \in \mathbb{F}^n$ associated to each eigenvalue λ by looking at solutions to the matrix equation $(A - \lambda I)v = 0$, where I denotes the identity map on $\mathbb{F}^n$.

(a) $\begin{bmatrix} -1 & 6 \\ 0 & 5 \end{bmatrix}$,   (b) $\begin{bmatrix} -\frac{1}{3} & 0 & 0 & 0 \\ 0 & -\frac{1}{3} & 0 & 0 \\ 0 & 0 & 1 & 0 \\ 0 & 0 & 0 & \frac{1}{2} \end{bmatrix}$,   (c) $\begin{bmatrix} 1 & 3 & 7 & 11 \\ 0 & \frac{1}{2} & 3 & 8 \\ 0 & 0 & 0 & 4 \\ 0 & 0 & 0 & 2 \end{bmatrix}$

(6) For each matrix A below, describe the invariant subspaces for the induced linear operator T on $\mathbb{F}^2$ that maps each $v \in \mathbb{F}^2$ to $T(v) = Av$.

(a) $\begin{bmatrix} 4 & -1 \\ 2 & 1 \end{bmatrix}$,   (b) $\begin{bmatrix} 0 & 1 \\ -1 & 0 \end{bmatrix}$,   (c) $\begin{bmatrix} 2 & 3 \\ 0 & 2 \end{bmatrix}$,   (d) $\begin{bmatrix} 1 & 0 \\ 0 & 0 \end{bmatrix}$

(7) Let $T \in \mathcal{L}(\mathbb{R}^2)$ be defined by

$$T\begin{pmatrix} x \\ y \end{pmatrix} = \begin{pmatrix} y \\ x + y \end{pmatrix}, \quad \text{for all } \begin{pmatrix} x \\ y \end{pmatrix} \in \mathbb{R}^2.$$

Define two real numbers λ_+ and λ_- as follows:

$$\lambda_+ = \frac{1 + \sqrt{5}}{2}, \quad \lambda_- = \frac{1 - \sqrt{5}}{2}.$$

  (a) Find the matrix of T with respect to the canonical basis for $\mathbb{R}^2$ (both as the domain and the codomain of T; call this matrix A).

  (b) Verify that λ_+ and λ_- are eigenvalues of T by showing that v_+ and v_- are eigenvectors, where

$$v_+ = \begin{pmatrix} 1 \\ \lambda_+ \end{pmatrix}, \quad v_- = \begin{pmatrix} 1 \\ \lambda_- \end{pmatrix}.$$

  (c) Show that (v_+, v_-) is a basis of $\mathbb{R}^2$.

  (d) Find the matrix of T with respect to the basis (v_+, v_-) for $\mathbb{R}^2$ (both as the domain and the codomain of T; call this matrix B).

Proof-Writing Exercises

(1) Let V be a finite-dimensional vector space over $\mathbb{F}$ with $T \in \mathcal{L}(V,V)$, and let $U_1, \ldots, U_m$ be subspaces of V that are invariant under T. Prove that $U_1 + \cdots + U_m$ must then also be an invariant subspace of V under T.

(2) Let V be a finite-dimensional vector space over $\mathbb{F}$ with $T \in \mathcal{L}(V,V)$, and suppose that U_1 and U_2 are subspaces of V that are invariant under T. Prove that $U_1 \cap U_2$ is also an invariant subspace of V under T.

(3) Let V be a finite-dimensional vector space over $\mathbb{F}$ with $T \in \mathcal{L}(V,V)$ invertible and $\lambda \in \mathbb{F} \setminus \{0\}$. Prove that λ is an eigenvalue for T if and only if λ^{-1} is an eigenvalue for T^{-1}.

(4) Let V be a finite-dimensional vector space over $\mathbb{F}$, and suppose that $T \in \mathcal{L}(V,V)$ has the property that every $v \in V$ is an eigenvector for T. Prove that T must then be a scalar multiple of the identity function on V.

(5) Let V be a finite-dimensional vector space over $\mathbb{F}$, and let $S, T \in \mathcal{L}(V)$ be linear operators on V with S invertible. Given any polynomial $p(z) \in \mathbb{F}[z]$, prove that

$$p(S \circ T \circ S^{-1}) = S \circ p(T) \circ S^{-1}.$$

(6) Let V be a finite-dimensional vector space over $\mathbb{C}$, $T \in \mathcal{L}(V)$ be a linear operator on V, and $p(z) \in \mathbb{C}[z]$ be a polynomial. Prove that $\lambda \in \mathbb{C}$ is an eigenvalue of the linear operator $p(T) \in \mathcal{L}(V)$ if and only if T has an eigenvalue $\mu \in \mathbb{C}$ such that $p(\mu) = \lambda$.

(7) Let V be a finite-dimensional vector space over $\mathbb{C}$ with $T \in \mathcal{L}(V)$ a linear operator on V. Prove that, for each $k = 1, \ldots, \dim(V)$, there is an invariant subspace U_k of V under T such that $\dim(U_k) = k$.

(8) Prove or give a counterexample to the following claim:

Claim. *Let V be a finite-dimensional vector space over $\mathbb{F}$, and let $T \in \mathcal{L}(V)$ be a linear operator on V. If the matrix for T with respect to some basis on V has all zeroes on the diagonal, then T is not invertible.*

(9) Prove or give a counterexample to the following claim:

Claim. *Let V be a finite-dimensional vector space over $\mathbb{F}$, and let $T \in \mathcal{L}(V)$ be a linear operator on V. If the matrix for T with respect to some basis on V has all non-zero elements on the diagonal, then T is invertible.*

(10) Let V be a finite-dimensional vector space over $\mathbb{F}$, and let $S, T \in \mathcal{L}(V)$ be linear operators on V. Suppose that T has $\dim(V)$ distinct eigenvalues and that, given any eigenvector $v \in V$ for T associated to some eigenvalue $\lambda \in \mathbb{F}$, v is also an eigenvector for S associated to some (possibly distinct) eigenvalue $\mu \in \mathbb{F}$. Prove that $T \circ S = S \circ T$.

(11) Let V be a finite-dimensional vector space over $\mathbb{F}$, and suppose that the linear operator $P \in \mathcal{L}(V)$ has the property that $P^2 = P$. Prove that $V = \text{null}(P) \oplus \text{range}(P)$.

(12) (a) Let $a, b, c, d \in \mathbb{F}$ and consider the system of equations given by

$$ax_1 + bx_2 = 0 \tag{7.4}$$

$$cx_1 + dx_2 = 0. \tag{7.5}$$

Note that $x_1 = x_2 = 0$ is a solution for any choice of a, b, c, and d. Prove that this system of equations has a non-trivial solution if and only if $ad - bc = 0$.

(b) Let $A = \begin{bmatrix} a & b \\ c & d \end{bmatrix} \in \mathbb{F}^{2 \times 2}$, and recall that we can define a linear operator $T \in \mathcal{L}(\mathbb{F}^2)$ on $\mathbb{F}^2$ by setting $T(v) = Av$ for each $v = \begin{bmatrix} v_1 \\ v_2 \end{bmatrix} \in \mathbb{F}^2$.

Show that the eigenvalues for T are exactly the $\lambda \in \mathbb{F}$ for which $p(\lambda) = 0$, where $p(z) = (a - z)(d - z) - bc$.

Hint: Write the eigenvalue equation $Av = \lambda v$ as $(A - \lambda I)v = 0$ and use the first part.

Chapter 8

Permutations and the Determinant of a Square Matrix

This chapter is devoted to an important quantity, called the **determinant**, which can be associated with any square matrix. In order to define the determinant, we will first need to define permutations.

8.1 Permutations

The study of permutations is a topic of independent interest with applications in many branches of mathematics such as Combinatorics and Probability Theory.

8.1.1 *Definition of permutations*

Given a positive integer $n \in \mathbb{Z}_+$, a **permutation** of an (ordered) list of n distinct objects is any reordering of this list. A permutation refers to the reordering itself and the nature of the objects involved is irrelevant. E.g., we can imagine interchanging the second and third items in a list of five distinct objects — no matter what those items are — and this defines a particular permutation that can be applied to any list of five objects.

Since the nature of the objects being rearranged (i.e., permuted) is immaterial, it is common to use the integers $1, 2, \ldots, n$ as the standard list of n objects. Alternatively, one can also think of these integers as labels for the items in any list of n distinct elements. This gives rise to the following definition.

Definition 8.1.1. A **permutation** π of n elements is a one-to-one and onto function having the set $\{1, 2, \ldots, n\}$ as both its domain and codomain.

In other words, a permutation is a function $\pi : \{1, 2, \ldots, n\} \longrightarrow \{1, 2, \ldots, n\}$ such that, for every integer $i \in \{1, \ldots, n\}$, there exists exactly one integer $j \in \{1, \ldots, n\}$ for which $\pi(j) = i$. We will usually denote permutations by Greek letters such as π (pi), σ (sigma), and τ (tau). The set of all permutations of n elements is denoted by $\mathcal{S}_n$ and is typically referred to as the **symmetric group of degree** n. (In particular, the set $\mathcal{S}_n$ forms a group under function composition as discussed in Section 8.1.2.)

Given a permutation $\pi \in \mathcal{S}_n$, there are several common notations used for specifying how π permutes the integers $1, 2, \ldots, n$.

Definition 8.1.2. Given a **permutation** $\pi \in \mathcal{S}_n$, denote $\pi_i = \pi(i)$ for each $i \in \{1, \ldots, n\}$. Then the **two-line notation** for π is given by the $2 \times n$ matrix

$$\pi = \begin{pmatrix} 1 & 2 & \cdots & n \\ \pi_1 & \pi_2 & \cdots & \pi_n \end{pmatrix}.$$

In other words, given a permutation $\pi \in \mathcal{S}_n$ and an integer $i \in \{1, \ldots, n\}$, we are denoting the image of i under π by π_i instead of using the more conventional function notation $\pi(i)$. Then, in order to specify the image of each integer $i \in \{1, \ldots, n\}$ under π, we list these images in a two-line array as shown above. (One can also use the so-called **one-line notation** for π, which is given by simply ignoring the top row and writing $\pi = \pi_1 \pi_2 \cdots \pi_n$.)

It is important to note that, although we represent permutations as $2 \times n$ matrices, you *should not* think of permutations as linear transformations from an n-dimensional vector space into a two-dimensional vector space. Moreover, the composition operation on permutation that we describe in Section 8.1.2 below *does not* correspond to matrix multiplication. The use of matrix notation in denoting permutations is merely a matter of convenience.

Example 8.1.3. Suppose that we have a set of five distinct objects and that we wish to describe the permutation that places the first item into the second position, the second item into the fifth position, the third item into the first position, the fourth item into the third position, and the fifth item into the fourth position. Then, using the notation developed above, we have the permutation $\pi \in \mathcal{S}_5$ such that

$$\pi_1 = \pi(1) = 3, \quad \pi_2 = \pi(2) = 1, \quad \pi_3 = \pi(3) = 4, \quad \pi_4 = \pi(4) = 5, \quad \pi_5 = \pi(5) = 2.$$

In two-line notation, we would write π as

$$\pi = \begin{pmatrix} 1\,2\,3\,4\,5 \\ 3\,1\,4\,5\,2 \end{pmatrix}.$$

It is relatively straightforward to find the number of permutations of n elements, i.e., to determine cardinality of the set $\mathcal{S}_n$. To construct an arbitrary permutation of n elements, we can proceed as follows: First, choose an integer $i \in \{1, \ldots, n\}$ to put into the first position. Clearly, we have exactly n possible choices. Next, choose the element to go in the second position. Since we have already chosen one element from the set $\{1, \ldots, n\}$, there are now exactly $n - 1$ remaining choices. Proceeding in this way, we have $n - 2$ choices when choosing the third element from the set $\{1, \ldots, n\}$, then $n - 3$ choices when choosing the fourth element, and so on until we are left with exactly one choice for the nth element. This proves the following theorem.

Theorem 8.1.4. *The number of elements in the symmetric group $\mathcal{S}_n$ is given by*

$$|\mathcal{S}_n| = n \cdot (n-1) \cdot (n-2) \cdot \cdots \cdot 3 \cdot 2 \cdot 1 = n!$$

We conclude this section with several examples, including a complete description of the one permutation in $\mathcal{S}_1$, the two permutations in $\mathcal{S}_2$, and the six permutations in $\mathcal{S}_3$. If you are patient you can list the $4! = 24$ permutations in $\mathcal{S}_4$ as further practice.

Example 8.1.5.

(1) Given any positive integer $n \in \mathbb{Z}_+$, the identity function $\text{id} : \{1, \ldots, n\} \longrightarrow \{1, \ldots, n\}$ given by $\text{id}(i) = i$, $\forall\, i \in \{1, \ldots, n\}$, is a permutation in $\mathcal{S}_n$. This function can be thought of as the trivial reordering that does not change the order at all, and so we call it the **trivial** or **identity** permutation.

(2) If $n = 1$, then, by Theorem 8.1.4, $|\mathcal{S}_n| = 1! = 1$. Thus, $\mathcal{S}_1$ contains only the identity permutation.

(3) If $n = 2$, then, by Theorem 8.1.4, $|\mathcal{S}_n| = 2! = 2 \cdot 1 = 2$. Thus, there is only one non-trivial permutation π in $\mathcal{S}_2$, namely the transformation interchanging the first and the second elements in a list. As a function, $\pi(1) = 2$ and $\pi(2) = 1$, and, in two-line notation,

$$\pi = \begin{pmatrix} 1 & 2 \\ \pi_1 & \pi_2 \end{pmatrix} = \begin{pmatrix} 1\,2 \\ 2\,1 \end{pmatrix}.$$

(4) If $n = 3$, then, by Theorem 8.1.4, $|\mathcal{S}_n| = 3! = 3 \cdot 2 \cdot 1 = 6$. Thus, there are five non-trivial permutations in $\mathcal{S}_3$. Using two-line notation, we have that

$$\mathcal{S}_3 = \left\{ \begin{pmatrix} 1\,2\,3 \\ 1\,2\,3 \end{pmatrix}, \begin{pmatrix} 1\,2\,3 \\ 1\,3\,2 \end{pmatrix}, \begin{pmatrix} 1\,2\,3 \\ 2\,1\,3 \end{pmatrix}, \begin{pmatrix} 1\,2\,3 \\ 2\,3\,1 \end{pmatrix}, \begin{pmatrix} 1\,2\,3 \\ 3\,1\,2 \end{pmatrix}, \begin{pmatrix} 1\,2\,3 \\ 3\,2\,1 \end{pmatrix} \right\}.$$

Keep in mind the fact that each element in $\mathcal{S}_3$ is simultaneously both a function and a reordering operation. E.g., the permutation

$$\pi = \begin{pmatrix} 1 & 2 & 3 \\ \pi_1 & \pi_2 & \pi_3 \end{pmatrix} = \begin{pmatrix} 1\,2\,3 \\ 2\,3\,1 \end{pmatrix}$$

can be read as defining the reordering that, with respect to the original list, places the second element in the first position, the third element in the second position, and the first element in the third position. This permutation could equally well have been identified by describing its action on the (ordered) list of letters a, b, c. In other words,

$$\begin{pmatrix} 1\,2\,3 \\ 2\,3\,1 \end{pmatrix} = \begin{pmatrix} a\,b\,c \\ b\,c\,a \end{pmatrix},$$

regardless of what the letters a, b, c might happen to represent.

8.1.2   *Composition of permutations*

Let $n \in \mathbb{Z}_+$ be a positive integer and $\pi, \sigma \in \mathcal{S}_n$ be permutations. Then, since π and σ are both functions from the set $\{1, \ldots, n\}$ to itself, we can compose them to obtain a new function $\pi \circ \sigma$ (read as *"pi after sigma"*) that takes on the values

$$(\pi \circ \sigma)(1) = \pi(\sigma(1)), \quad (\pi \circ \sigma)(2) = \pi(\sigma(2)), \quad \ldots \quad (\pi \circ \sigma)(n) = \pi(\sigma(n)).$$

In two-line notation, we can write $\pi \circ \sigma$ as

$$\begin{pmatrix} 1 & 2 & \cdots & n \\ \pi(\sigma(1)) & \pi(\sigma(2)) & \cdots & \pi(\sigma(n)) \end{pmatrix} \text{ or } \begin{pmatrix} 1 & 2 & \cdots & n \\ \pi_{\sigma(1)} & \pi_{\sigma(2)} & \cdots & \pi_{\sigma(n)} \end{pmatrix} \text{ or } \begin{pmatrix} 1 & 2 & \cdots & n \\ \pi_{\sigma_1} & \pi_{\sigma_2} & \cdots & \pi_{\sigma_n} \end{pmatrix}.$$

Example 8.1.6. From $\mathcal{S}_3$, suppose that we have the permutations π and σ given by

$$\pi(1) = 2, \ \pi(2) = 3, \ \pi(3) = 1 \ \text{ and } \ \sigma(1) = 1, \ \sigma(2) = 3, \ \sigma(3) = 2.$$

Then note that

$$(\pi \circ \sigma)(1) = \pi(\sigma(1)) = \pi(1) = 2,$$

$$(\pi \circ \sigma)(2) = \pi(\sigma(2)) = \pi(3) = 1,$$

$$(\pi \circ \sigma)(3) = \pi(\sigma(3)) = \pi(2) = 3.$$

In other words,

$$\begin{pmatrix} 1\,2\,3 \\ 2\,3\,1 \end{pmatrix} \circ \begin{pmatrix} 1\,2\,3 \\ 1\,3\,2 \end{pmatrix} = \begin{pmatrix} 1 & 2 & 3 \\ \pi(1) & \pi(3) & \pi(2) \end{pmatrix} = \begin{pmatrix} 1\,2\,3 \\ 2\,1\,3 \end{pmatrix}.$$

Similar computations (which you should check for your own practice) yield compositions such as

$$\begin{pmatrix} 1\,2\,3 \\ 1\,3\,2 \end{pmatrix} \circ \begin{pmatrix} 1\,2\,3 \\ 2\,3\,1 \end{pmatrix} = \begin{pmatrix} 1 & 2 & 3 \\ \sigma(2) & \sigma(3) & \sigma(1) \end{pmatrix} = \begin{pmatrix} 1\,2\,3 \\ 3\,2\,1 \end{pmatrix},$$

$$\begin{pmatrix} 1\,2\,3 \\ 2\,3\,1 \end{pmatrix} \circ \begin{pmatrix} 1\,2\,3 \\ 1\,2\,3 \end{pmatrix} = \begin{pmatrix} 1 & 2 & 3 \\ \sigma(1) & \sigma(2) & \sigma(3) \end{pmatrix} = \begin{pmatrix} 1\,2\,3 \\ 2\,3\,1 \end{pmatrix},$$

and

$$\begin{pmatrix} 1\,2\,3 \\ 1\,2\,3 \end{pmatrix} \circ \begin{pmatrix} 1\,2\,3 \\ 2\,3\,1 \end{pmatrix} = \begin{pmatrix} 1 & 2 & 3 \\ \text{id}(2) & \text{id}(3) & \text{id}(1) \end{pmatrix} = \begin{pmatrix} 1\,2\,3 \\ 2\,3\,1 \end{pmatrix}.$$

In particular, note that the result of each composition above is a permutation, that composition is not a commutative operation, and that composition with id leaves a permutation unchanged. Moreover, since each permutation π is a bijection, one can always construct an inverse permutation π^{-1} such that $\pi \circ \pi^{-1} = \text{id}$. E.g.,

$$\begin{pmatrix} 1\,2\,3 \\ 2\,3\,1 \end{pmatrix} \circ \begin{pmatrix} 1\,2\,3 \\ 3\,1\,2 \end{pmatrix} = \begin{pmatrix} 1 & 2 & 3 \\ \pi(3) & \pi(1) & \pi(2) \end{pmatrix} = \begin{pmatrix} 1\,2\,3 \\ 1\,2\,3 \end{pmatrix}.$$

We summarize the basic properties of composition on the symmetric group in the following theorem.

Theorem 8.1.7. *Let $n \in \mathbb{Z}_+$ be a positive integer. Then the set $\mathcal{S}_n$ has the following properties.*

(1) Given any two permutations $\pi, \sigma \in \mathcal{S}_n$, the composition $\pi \circ \sigma \in \mathcal{S}_n$.

(2) (Associativity of Composition) Given any three permutations $\pi, \sigma, \tau \in \mathcal{S}_n$,

$$(\pi \circ \sigma) \circ \tau = \pi \circ (\sigma \circ \tau).$$

(3) (Identity Element for Composition) Given any permutation $\pi \in \mathcal{S}_n$,

$$\pi \circ id = id \circ \pi = \pi.$$

(4) (Inverse Elements for Composition) Given any permutation $\pi \in \mathcal{S}_n$, there exists a unique permutation $\pi^{-1} \in \mathcal{S}_n$ such that

$$\pi \circ \pi^{-1} = \pi^{-1} \circ \pi = id.$$

In other words, the set $\mathcal{S}_n$ forms a group under composition.

Note that the composition of permutations is not commutative in general. In particular, for $n \geq 3$, it is easy to find permutations π and σ such that $\pi \circ \sigma \neq \sigma \circ \pi$.

8.1.3 *Inversions and the sign of a permutation*

Let $n \in \mathbb{Z}_+$ be a positive integer. Then, given a permutation $\pi \in \mathcal{S}_n$, it is natural to ask how "out of order" π is in comparison to the identity permutation. One method for quantifying this is to count the number of so-called **inversion pairs** in π as these describe pairs of objects that are out of order relative to each other.

Definition 8.1.8. Let $\pi \in \mathcal{S}_n$ be a permutation. Then an **inversion pair** (i, j) of π is a pair of positive integers $i, j \in \{1, \ldots, n\}$ for which $i < j$ but $\pi(i) > \pi(j)$.

Note, in particular, that the components of an inversion pair are the **positions** where the two "out of order" elements occur. An inversion pair is often referred to simply as an inversion.

Example 8.1.9. We classify all inversion pairs for elements in $\mathcal{S}_3$:

- $id = \begin{pmatrix} 1\ 2\ 3 \\ 1\ 2\ 3 \end{pmatrix}$ has no inversion pairs since no elements are "out of order".

- $\pi = \begin{pmatrix} 1\ 2\ 3 \\ 1\ 3\ 2 \end{pmatrix}$ has the single inversion pair $(2, 3)$ since $\pi(2) = 3 > 2 = \pi(3)$.

- $\pi = \begin{pmatrix} 1\ 2\ 3 \\ 2\ 1\ 3 \end{pmatrix}$ has the single inversion pair $(1, 2)$ since $\pi(1) = 2 > 1 = \pi(2)$.

- $\pi = \begin{pmatrix} 1\ 2\ 3 \\ 2\ 3\ 1 \end{pmatrix}$ has the two inversion pairs $(1, 3)$ and $(2, 3)$ since we have that both $\pi(1) = 2 > 1 = \pi(3)$ and $\pi(2) = 3 > 1 = \pi(3)$.

- $\pi = \begin{pmatrix} 1\ 2\ 3 \\ 3\ 1\ 2 \end{pmatrix}$ has the two inversion pairs $(1,2)$ and $(1,3)$ since we have that both $\pi(1) = 3 > 1 = \pi(2)$ and $\pi(1) = 3 > 2 = \pi(3)$.

- $\pi = \begin{pmatrix} 1\ 2\ 3 \\ 3\ 2\ 1 \end{pmatrix}$ has the three inversion pairs $(1,2)$, $(1,3)$, and $(2,3)$, as you can check.

Example 8.1.10. As another example, for each $i, j \in \{1, \ldots, n\}$ with $i < j$, we define the **transposition** $t_{ij} \in \mathcal{S}_n$ by

$$t_{ij} = \begin{pmatrix} 1\ 2\ \cdots\ i\ \cdots\ j\ \cdots\ n \\ 1\ 2\ \cdots\ j\ \cdots\ i\ \cdots\ n \end{pmatrix}.$$

In other words, t_{ij} is the permutation that interchanges i and j while leaving all other integers fixed in place. One can check that the number of inversions in t_{ij} is exactly $2(j - i) - 1$. Thus, the number of inversions in a transposition is always odd. E.g.,

$$t_{13} = \begin{pmatrix} 1\ 2\ 3\ 4 \\ 3\ 2\ 1\ 4 \end{pmatrix}$$

has inversion pairs $(1,2)$, $(1,3)$, and $(2,3)$.

For our purposes in this text, the significance of inversion pairs is mainly due to the following fundamental definition.

Definition 8.1.11. Let $\pi \in \mathcal{S}_n$ be a permutation. Then the **sign** of π, denoted by $\mathrm{sign}(\pi)$, is defined by

$$\mathrm{sign}(\pi) = (-1)^{\#\ \text{of inversion pairs in}\ \pi} = \begin{cases} +1, & \text{if the number of inversions in } \pi \text{ is even} \\ -1, & \text{if the number of inversions in } \pi \text{ is odd} \end{cases}.$$

We call π an **even permutation** if $\mathrm{sign}(\pi) = +1$, whereas π is called an **odd permutation** if $\mathrm{sign}(\pi) = -1$.

Example 8.1.12. Based upon the computations in Example 8.1.9 above, we have that

$$\mathrm{sign}\begin{pmatrix} 1\ 2\ 3 \\ 1\ 2\ 3 \end{pmatrix} = \mathrm{sign}\begin{pmatrix} 1\ 2\ 3 \\ 2\ 3\ 1 \end{pmatrix} = \mathrm{sign}\begin{pmatrix} 1\ 2\ 3 \\ 3\ 1\ 2 \end{pmatrix} = +1$$

and that

$$\mathrm{sign}\begin{pmatrix} 1\ 2\ 3 \\ 1\ 3\ 2 \end{pmatrix} = \mathrm{sign}\begin{pmatrix} 1\ 2\ 3 \\ 2\ 1\ 3 \end{pmatrix} = \mathrm{sign}\begin{pmatrix} 1\ 2\ 3 \\ 3\ 2\ 1 \end{pmatrix} = -1.$$

Similarly, from Example 8.1.10, it follows that any transposition is an odd permutation.

We summarize some of the most basic properties of the sign operation on the symmetric group in the following theorem.

Theorem 8.1.13. *Let $n \in \mathbb{Z}_+$ be a positive integer. Then,*

(1) for $id \in S_n$ the identity permutation,
$$sign(id) = +1.$$

(2) for $t_{ij} \in S_n$ a transposition with $i, j \in \{1, \ldots, n\}$ and $i < j$,
$$sign(t_{ij}) = -1. \tag{8.1}$$

(3) given any two permutations $\pi, \sigma \in S_n$,
$$sign(\pi \circ \sigma) = sign(\pi)\, sign(\sigma), \tag{8.2}$$
$$sign(\pi^{-1}) = sign(\pi). \tag{8.3}$$

(4) the number of even permutations in S_n, when $n \geq 2$, is exactly $\frac{1}{2}n!$.
(5) the set A_n of even permutations in S_n forms a group under composition.

8.2 Determinants

Now that we have developed the appropriate background material on permutations, we are finally ready to define the determinant and explore its many important properties.

8.2.1 *Summations indexed by the set of all permutations*

Given a positive integer $n \in \mathbb{Z}_+$, we begin with the following definition:

Definition 8.2.1. Given a square matrix $A = (a_{ij}) \in \mathbb{F}^{n \times n}$, the **determinant** of A is defined to be

$$\det(A) = \sum_{\pi \in S_n} sign(\pi) a_{1,\pi(1)} a_{2,\pi(2)} \cdots a_{n,\pi(n)}, \tag{8.4}$$

where the sum is over all permutations of n elements (i.e., over the symmetric group).

Note that each permutation in the summand of (8.4) permutes the n columns of the $n \times n$ matrix.

Example 8.2.2. Suppose that $A \in \mathbb{F}^{2 \times 2}$ is the 2×2 matrix

$$A = \begin{bmatrix} a_{11} & a_{12} \\ a_{21} & a_{22} \end{bmatrix}.$$

To calculate the determinant of A, we first list the two permutations in S_2:

$$id = \begin{pmatrix} 1 & 2 \\ 1 & 2 \end{pmatrix} \qquad \text{and} \qquad \sigma = \begin{pmatrix} 1 & 2 \\ 2 & 1 \end{pmatrix}.$$

The permutation id has sign 1, and the permutation σ has sign -1. Thus, the determinant of A is given by

$$\det(A) = a_{11}a_{22} - a_{12}a_{21}.$$

If one attempted to compute determinants directly using Equation (8.4), then one would need to sum up $n!$ terms, where each summand is itself a product of n factors. This is an incredibly inefficient method for finding determinants since $n!$ increases in size very rapidly as n increases. E.g., $10! = 3628800$. Thus, even if you could compute one summand per second without stopping, it would still take you well over a month to compute the determinant of a 10×10 matrix using Equation (8.4). Fortunately, there are properties of the determinant (as summarized in Section 8.2.2 below) that can be used to greatly reduce the size of such computations. These properties of the determinant follow from general properties that hold for any summation taken over the symmetric group, which are in turn themselves based upon properties of permutations and the fact that addition and multiplication are commutative operations in the field $\mathbb{F}$ (which, as usual, we take to be either $\mathbb{R}$ or $\mathbb{C}$).

Let $T : \mathcal{S}_n \to V$ be a function defined on the symmetric group $\mathcal{S}_n$ that takes values in some vector space V. E.g., $T(\pi)$ could be the term corresponding to the permutation π in Equation (8.4). Then, since the sum

$$\sum_{\pi \in \mathcal{S}_n} T(\pi)$$

is finite, we are free to reorder the summands. In other words, the sum is independent of the order in which the terms are added, and so we are free to permute the term order without affecting the value of the sum. Some commonly used reorderings of such sums are the following:

$$\sum_{\pi \in \mathcal{S}_n} T(\pi) = \sum_{\pi \in \mathcal{S}_n} T(\sigma \circ \pi) \tag{8.5}$$

$$= \sum_{\pi \in \mathcal{S}_n} T(\pi \circ \sigma) \tag{8.6}$$

$$= \sum_{\pi \in \mathcal{S}_n} T(\pi^{-1}), \tag{8.7}$$

where σ is a fixed permutation.

Equation (8.5) follows from the fact that, if π runs through each permutation in $\mathcal{S}_n$ exactly once, then $\sigma \circ \pi$ similarly runs through each permutation but in a potentially different order. I.e., the action of σ upon something like Equation (8.4) is that σ merely permutes the permutations that index the terms. Put another way, there is a one-to-one correspondence between permutations in general and permutations composed with σ.

Similar reasoning holds for Equations (8.6) and (8.7).

8.2.2 *Properties of the determinant*

We summarize some of the most basic properties of the determinant below. The proof of the following theorem uses properties of permutations, properties of the sign function on permutations, and properties of sums over the symmetric group as discussed in Section 8.2.1 above. In thinking about these properties, it is useful to keep in mind that, using Equation (8.4), the determinant of an $n \times n$ matrix A is the sum over all possible ways of selecting n entries of A, where exactly one element is selected from each row and from each column of A.

Theorem 8.2.3 (Properties of the Determinant). *Let $n \in \mathbb{Z}_+$ be a positive integer, and suppose that $A = (a_{ij}) \in \mathbb{F}^{n \times n}$ is an $n \times n$ matrix. Then*

(1) $\det(0_{n \times n}) = 0$ *and* $\det(I_n) = 1$, *where* $0_{n \times n}$ *denotes the* $n \times n$ *zero matrix and* I_n *denotes the* $n \times n$ *identity matrix.*

(2) $\det(A^T) = \det(A)$, *where* A^T *denotes the transpose of A.*

(3) denoting by $A^{(\cdot,1)}, A^{(\cdot,2)}, \ldots, A^{(\cdot,n)} \in \mathbb{F}^n$ *the columns of A, $\det(A)$ is a linear function of column $A^{(\cdot,i)}$, for each $i \in \{1, \ldots, n\}$. In other words, if we denote*

$$A = \left[A^{(\cdot,1)} \mid A^{(\cdot,2)} \mid \cdots \mid A^{(\cdot,n)} \right]$$

then, given any scalar $z \in \mathbb{F}$ and any vectors $a_1, a_2, \ldots, a_n, c, b \in \mathbb{F}^n$,

$$\det [a_1 \mid \cdots \mid a_{i-1} \mid za_i \mid \cdots \mid a_n] = z \det [a_1 \mid \cdots \mid a_{i-1} \mid a_i \mid \cdots \mid a_n],$$

$$\det [a_1 \mid \cdots \mid a_{i-1} \mid b + c \mid \cdots \mid a_n] = \det [a_1 \mid \cdots \mid b \mid \cdots \mid a_n]$$
$$+ \det [a_1 \mid \cdots \mid c \mid \cdots \mid a_n].$$

(4) $\det(A)$ is an antisymmetric function of the columns of A. In other words, given any positive integers $1 \leq i < j \leq n$ and denoting $A = \left[A^{(\cdot,1)} \mid A^{(\cdot,2)} \mid \cdots \mid A^{(\cdot,n)} \right]$,

$$\det(A) = - \det \left[A^{(\cdot,1)} \mid \cdots \mid A^{(\cdot,j)} \mid \cdots \mid A^{(\cdot,i)} \mid \cdots \mid A^{(\cdot,n)} \right].$$

(5) if A has two identical columns, $\det(A) = 0$.

(6) if A has a column of zeroes, $\det(A) = 0$.

(7) Properties (3)–(6) also hold when rows are used in place of columns.

(8) given any other matrix $B \in \mathbb{F}^{n \times n}$,

$$\det(AB) = \det(A) \det(B).$$

(9) if A is either upper triangular or lower triangular,

$$\det(A) = a_{11} a_{22} \cdots a_{nn}.$$

Proof. First, note that Properties (1), (3), (6), and (9) follow directly from the sum given in Equation (8.4). Moreover, Property (5) follows directly from Property (4), and Property (7) follows directly from Property (2). Thus, we only need to prove Properties (2), (4), and (8).

Proof of (2). Since the entries of A^T are obtained from those of A by interchanging the row and column indices, it follows that $\det(A^T)$ is given by

$$\det(A^T) = \sum_{\pi \in \mathcal{S}_n} \text{sign}(\pi) \, a_{\pi(1),1} a_{\pi(2),2} \cdots a_{\pi(n),n} \, .$$

Using the commutativity of the product in $\mathbb{F}$ and Equation (8.3), we see that

$$\det(A^T) = \sum_{\pi \in \mathcal{S}_n} \text{sign}(\pi^{-1})\, a_{1,\pi^{-1}(1)} a_{2,\pi^{-1}(2)} \cdots a_{n,\pi^{-1}(n)},$$

which equals $\det(A)$ by Equation (8.7).

Proof of (4). Let $B = \left[A^{(\cdot,1)} \mid \cdots \mid A^{(\cdot,j)} \mid \cdots \mid A^{(\cdot,i)} \mid \cdots \mid A^{(\cdot,n)}\right]$ be the matrix obtained from A by interchanging the ith and the jth column. Then note that

$$\det(B) = \sum_{\pi \in \mathcal{S}_n} \text{sign}(\pi)\, a_{1,\pi(1)} \cdots a_{j,\pi(i)} \cdots a_{i,\pi(j)} \cdots a_{n,\pi(n)}.$$

Define $\tilde{\pi} = \pi \circ t_{ij}$, and note that $\pi = \tilde{\pi} \circ t_{ij}$. In particular, $\pi(i) = \tilde{\pi}(j)$ and $\pi(j) = \tilde{\pi}(i)$, from which

$$\det(B) = \sum_{\pi \in \mathcal{S}_n} \text{sign}(\tilde{\pi} \circ t_{ij})\, a_{1,\tilde{\pi}(1)} \cdots a_{i,\tilde{\pi}(i)} \cdots a_{j,\tilde{\pi}(j)} \cdots a_{n,\tilde{\pi}(n)}.$$

It follows from Equations (8.2) and (8.1) that $\text{sign}(\tilde{\pi} \circ t_{ij}) = -\text{sign}(\tilde{\pi})$. Thus, using Equation (8.6), we obtain $\det(B) = -\det(A)$.

Proof of (8). Using the standard expression for the matrix entries of the product AB in terms of the matrix entries of $A = (a_{ij})$ and $B = (b_{ij})$, we have that

$$\det(AB) = \sum_{\pi \in \mathcal{S}_n} \text{sign}(\pi) \sum_{k_1=1}^{n} \cdots \sum_{k_n=1}^{n} a_{1,k_1} b_{k_1,\pi(1)} \cdots a_{n,k_n} b_{k_n,\pi(n)}$$

$$= \sum_{k_1=1}^{n} \cdots \sum_{k_n=1}^{n} a_{1,k_1} \cdots a_{n,k_n} \sum_{\pi \in \mathcal{S}_n} \text{sign}(\pi) b_{k_1,\pi(1)} \cdots b_{k_n,\pi(n)}.$$

Note that, for fixed $k_1,\ldots,k_n \in \{1,\ldots,n\}$, the sum $\sum_{\pi \in \mathcal{S}_n} \text{sign}(\pi) b_{k_1,\pi(1)} \cdots b_{k_n,\pi(n)}$ is the determinant of a matrix composed of rows $k_1,\ldots,k_n$ of B. Thus, by Property (5), it follows that this expression vanishes unless the k_i are pairwise distinct. In other words, the sum over all choices of $k_1,\ldots,k_n$ can be restricted to those sets of indices $\sigma(1),\ldots,\sigma(n)$ that are labeled by a permutation $\sigma \in \mathcal{S}_n$. In other words,

$$\det(AB) = \sum_{\sigma \in \mathcal{S}_n} a_{1,\sigma(1)} \cdots a_{n,\sigma(n)} \sum_{\pi \in \mathcal{S}_n} \text{sign}(\pi)\, b_{\sigma(1),\pi(1)} \cdots b_{\sigma(n),\pi(n)}.$$

Now, proceeding with the same arguments as in the proof of Property (4) but with the role of t_{ij} replaced by an arbitrary permutation σ, we obtain

$$\det(AB) = \sum_{\sigma \in \mathcal{S}_n} \text{sign}(\sigma)\, a_{1,\sigma(1)} \cdots a_{n,\sigma(n)} \sum_{\pi \in \mathcal{S}_n} \text{sign}(\pi \circ \sigma^{-1})\, b_{1,\pi \circ \sigma^{-1}(1)} \cdots b_{n,\pi \circ \sigma^{-1}(n)}.$$

Using Equation (8.6), this last expression then becomes $(\det(A))(\det(B))$. $\qquad\square$

Note that Properties (3) and (4) of Theorem 8.2.3 effectively summarize how multiplication by an Elementary Matrix interacts with the determinant operation. These Properties together with Property (9) facilitate numerical computation of determinants of larger matrices.

8.2.3   *Further properties and applications*

There are many applications of Theorem 8.2.3. We conclude this chapter with a few consequences that are particularly useful when computing with matrices. In particular, we use the determinant to list several characterizations for matrix invertibility, and, as a corollary, give a method for using determinants to calculate eigenvalues. You should provide a proof of these results for your own benefit.

Theorem 8.2.4. *Let $n \in \mathbb{Z}_+$ and $A \in \mathbb{F}^{n \times n}$. Then the following statements are equivalent:*

(1) A is invertible.

(2) denoting $x = \begin{bmatrix} x_1 \\ \vdots \\ x_n \end{bmatrix}$, the matrix equation $Ax = 0$ has only the trivial solution $x = 0$.

(3) denoting $x = \begin{bmatrix} x_1 \\ \vdots \\ x_n \end{bmatrix}$, the matrix equation $Ax = b$ has a solution for all $b = \begin{bmatrix} b_1 \\ \vdots \\ b_n \end{bmatrix} \in \mathbb{F}^n$.

(4) A can be factored into a product of elementary matrices.
(5) $\det(A) \neq 0$.
(6) the rows (or columns) of A form a linearly independent set in $\mathbb{F}^n$.
(7) zero is not an eigenvalue of A.
(8) the linear operator $T : \mathbb{F}^n \to \mathbb{F}^n$ defined by $T(x) = Ax$, for every $x \in \mathbb{F}^n$, is bijective.

Moreover, should A be invertible, then $\det(A^{-1}) = \dfrac{1}{\det(A)}$.

Given a matrix $A \in \mathbb{C}^{n \times n}$ and a complex number $\lambda \in \mathbb{C}$, the expression

$$P(\lambda) = \det(A - \lambda I_n)$$

is called the **characteristic polynomial** of A. Note that $P(\lambda)$ is a basis independent polynomial of degree n. Thus, as with the determinant, we can consider $P(\lambda)$ to be associated with the linear map that has matrix A with respect to some basis. Since the eigenvalues of A are exactly those $\lambda \in \mathbb{C}$ such that $A - \lambda I$ is not invertible, the following is then an immediate corollary.

Corollary 8.2.5. *The roots of the polynomial $P(\lambda) = \det(A - \lambda I)$ are exactly the eigenvalues of A.*

8.2.4   *Computing determinants with cofactor expansions*

As noted in Section 8.2.1, it is generally impractical to compute determinants directly with Equation (8.4). In this section, we briefly describe the so-called *cofactor expansions* of a determinant. When properly applied, cofactor expansions are particularly useful for computing determinants by hand.

Definition 8.2.6. Let $n \in \mathbb{Z}_+$ and $A \in \mathbb{F}^{n \times n}$. Then, for each $i, j \in \{1, 2, \ldots, n\}$, the *i-j minor* of A, denoted M_{ij}, is defined to be the determinant of the matrix obtained by removing the ith row and j^{th} column from A. Moreover, the *i-j cofactor* of A is defined to be

$$A_{ij} = (-1)^{i+j} M_{ij}.$$

Cofactors themselves, though, are not terribly useful unless put together in the right way.

Definition 8.2.7. Let $n \in \mathbb{Z}_+$ and $A = (a_{ij}) \in \mathbb{F}^{n \times n}$. Then, for each $i, j \in \{1, 2, \ldots, n\}$, the ith *row* (resp. j^{th} *column*) *cofactor expansion* of A is the sum $\sum_{j=1}^{n} a_{ij} A_{ij}$ (resp. $\sum_{i=1}^{n} a_{ij} A_{ij}$).

Theorem 8.2.8. *Let $n \in \mathbb{Z}_+$ and $A \in \mathbb{F}^{n \times n}$. Then every row and column factor expansion of A is equal to the determinant of A.*

Since the determinant of a matrix is equal to every row or column cofactor expansion, one can compute the determinant using a convenient choice of expansions until the calculation is reduced to one or more 2×2 determinants. We close with an example.

Example 8.2.9. By first expanding along the second column, we obtain

$$\begin{vmatrix} 1 & 2 & -3 & 4 \\ -4 & 2 & 1 & 3 \\ 3 & 0 & 0 & -3 \\ 2 & 0 & -2 & 3 \end{vmatrix} = (-1)^{1+2}(2) \begin{vmatrix} -4 & 1 & 3 \\ 3 & 0 & -3 \\ 2 & -2 & 3 \end{vmatrix} + (-1)^{2+2}(2) \begin{vmatrix} 1 & -3 & 4 \\ 3 & 0 & -3 \\ 2 & -2 & 3 \end{vmatrix}.$$

Then, each of the resulting 3×3 determinants can be computed by further expansion:

$$\begin{vmatrix} -4 & 1 & 3 \\ 3 & 0 & -3 \\ 2 & -2 & 3 \end{vmatrix} = (-1)^{1+2}(1) \begin{vmatrix} 3 & -3 \\ 2 & 3 \end{vmatrix} + (-1)^{3+2}(-2) \begin{vmatrix} -4 & 3 \\ 3 & -3 \end{vmatrix} = -15 + 6 = -9.$$

$$\begin{vmatrix} 1 & -3 & 4 \\ 3 & 0 & -3 \\ 2 & -2 & 3 \end{vmatrix} = (-1)^{2+1}(3) \begin{vmatrix} -3 & 4 \\ -2 & 3 \end{vmatrix} + (-1)^{2+3}(-3) \begin{vmatrix} 1 & -3 \\ 2 & -2 \end{vmatrix} = 3 + 12 = 15.$$

It follows that the original determinant is then equal to $-2(-9) + 2(15) = 48$.

Exercises for Chapter 8

Calculational Exercises

(1) Let $A \in \mathbb{C}^{3 \times 3}$ be given by

$$A = \begin{bmatrix} 1 & 0 & i \\ 0 & 1 & 0 \\ -i & 0 & -1 \end{bmatrix}.$$

    (a) Calculate $\det(A)$.
    (b) Find $\det(A^4)$.

(2) (a) For each permutation $\pi \in S_3$, compute the number of inversions in π, and classify π as being either an even or an odd permutation.
    (b) Use your result from Part (a) to construct a formula for the determinant of a 3×3 matrix.

(3) (a) For each permutation $\pi \in S_4$, compute the number of inversions in π, and classify π as being either an even or an odd permutation.
    (b) Use your result from Part (a) to construct a formula for the determinant of a 4×4 matrix.

(4) Solve for the variable x in the following expression:

$$\det\left(\begin{bmatrix} x & -1 \\ 3 & 1-x \end{bmatrix} \right) = \det\left(\begin{bmatrix} 1 & 0 & -3 \\ 2 & x & -6 \\ 1 & 3 & x-5 \end{bmatrix} \right).$$

(5) Prove that the following determinant does not depend upon the value of θ:

$$\det\left(\begin{bmatrix} \sin(\theta) & \cos(\theta) & 0 \\ -\cos(\theta) & \sin(\theta) & 0 \\ \sin(\theta) - \cos(\theta) & \sin(\theta) + \cos(\theta) & 1 \end{bmatrix} \right).$$

(6) Given scalars $\alpha, \beta, \gamma \in \mathbb{F}$, prove that the following matrix is not invertible:

$$\begin{bmatrix} \sin^2(\alpha) & \sin^2(\beta) & \sin^2(\gamma) \\ \cos^2(\alpha) & \cos^2(\beta) & \cos^2(\gamma) \\ 1 & 1 & 1 \end{bmatrix}.$$

    Hint: Compute the determinant.

Proof-Writing Exercises

(1) Let $a, b, c, d, e, f \in \mathbb{F}$ be scalars, and suppose that A and B are the following matrices:

$$A = \begin{bmatrix} a & b \\ 0 & c \end{bmatrix} \quad \text{and} \quad B = \begin{bmatrix} d & e \\ 0 & f \end{bmatrix}.$$

Prove that $AB = BA$ if and only if $\det\left(\begin{bmatrix} b & a-c \\ e & d-f \end{bmatrix} \right) = 0$.

(2) Given a square matrix A, prove that A is invertible if and only if $A^T A$ is invertible.

(3) Prove or give a counterexample: For any $n \geq 1$ and $A, B \in \mathbb{R}^{n \times n}$, one has

$$\det(A + B) = \det(A) + \det(B).$$

(4) Prove or give a counterexample: For any $r \in \mathbb{R}$, $n \geq 1$ and $A \in \mathbb{R}^{n \times n}$, one has

$$\det(rA) = r \det(A).$$

Chapter 9

Inner Product Spaces

The abstract definition of a vector space only takes into account algebraic properties for the addition and scalar multiplication of vectors. For vectors in $\mathbb{R}^n$, for example, we also have geometric intuition involving the length of a vector or the angle formed by two vectors. In this chapter we discuss inner product spaces, which are vector spaces with an inner product defined upon them. Using the inner product, we will define notions such as the length of a vector, orthogonality, and the angle between non-zero vectors.

9.1 Inner product

In this section, V is a finite-dimensional, non-zero vector space over $\mathbb{F}$.

Definition 9.1.1. An **inner product** on V is a map

$$\langle \cdot, \cdot \rangle : V \times V \to \mathbb{F}$$
$$(u, v) \mapsto \langle u, v \rangle$$

with the following four properties.

(1) **Linearity in first slot:** $\langle u + v, w \rangle = \langle u, w \rangle + \langle v, w \rangle$ and $\langle au, v \rangle = a\langle u, v \rangle$ for all $u, v, w \in V$ and $a \in \mathbb{F}$;
(2) **Positivity:** $\langle v, v \rangle \geq 0$ for all $v \in V$;
(3) **Positive definiteness:** $\langle v, v \rangle = 0$ if and only if $v = 0$;
(4) **Conjugate symmetry:** $\langle u, v \rangle = \overline{\langle v, u \rangle}$ for all $u, v \in V$.

Remark 9.1.2. Recall that every real number $x \in \mathbb{R}$ equals its complex conjugate. Hence, for real vector spaces, conjugate symmetry of an inner product becomes actual symmetry.

Definition 9.1.3. An **inner product space** is a vector space over $\mathbb{F}$ together with an inner product $\langle \cdot, \cdot \rangle$.

Example 9.1.4. Let $V = \mathbb{F}^n$ and $u = (u_1, \ldots, u_n), v = (v_1, \ldots, v_n) \in \mathbb{F}^n$. Then we can define an inner product on V by setting

$$\langle u, v \rangle = \sum_{i=1}^{n} u_i \bar{v}_i.$$

For $\mathbb{F} = \mathbb{R}$, this reduces to the usual dot product, i.e.,

$$u \cdot v = u_1 v_1 + \cdots + u_n v_n.$$

Example 9.1.5. Let $V = \mathbb{F}[z]$ be the space of polynomials with coefficients in $\mathbb{F}$. Given $f, g \in \mathbb{F}[z]$, we can define their inner product to be

$$\langle f, g \rangle = \int_0^1 f(z)\overline{g(z)}dz,$$

where $\overline{g(z)}$ is the complex conjugate of the polynomial $g(z)$.

For a fixed vector $w \in V$, one can define a map $T : V \to \mathbb{F}$ by setting $Tv = \langle v, w \rangle$. Note that T is linear by Condition 1 of Definition 9.1.1. This implies, in particular, that $\langle 0, w \rangle = 0$ for every $w \in V$. By conjugate symmetry, we also have $\langle w, 0 \rangle = 0$.

Lemma 9.1.6. *The inner product is anti-linear in the second slot, that is, $\langle u, v + w \rangle = \langle u, v \rangle + \langle u, w \rangle$ and $\langle u, av \rangle = \bar{a}\langle u, v \rangle$ for all $u, v, w \in V$ and $a \in \mathbb{F}$.*

Proof. For additivity, note that

$$\langle u, v + w \rangle = \overline{\langle v + w, u \rangle} = \overline{\langle v, u \rangle + \langle w, u \rangle}$$
$$= \overline{\langle v, u \rangle} + \overline{\langle w, u \rangle} = \langle u, v \rangle + \langle u, w \rangle.$$

Similarly, for anti-homogeneity, note that

$$\langle u, av \rangle = \overline{\langle av, u \rangle} = \overline{a\langle v, u \rangle} = \bar{a}\overline{\langle v, u \rangle} = \bar{a}\langle u, v \rangle.$$

$\square$

We close this section by noting that the convention in physics is often the exact opposite of what we have defined above. In other words, an inner product in physics is traditionally linear in the second slot and anti-linear in the first slot.

9.2 Norms

The norm of a vector in an arbitrary inner product space is the analog of the length or magnitude of a vector in $\mathbb{R}^n$. We formally define this concept as follows.

Definition 9.2.1. Let V be a vector space over $\mathbb{F}$. A map

$$\| \cdot \| : V \to \mathbb{R}$$
$$v \mapsto \|v\|$$

is a **norm** on V if the following three conditions are satisfied.

(1) **Positive definiteness:** $\|v\| = 0$ if and only if $v = 0$;
(2) **Positive homogeneity:** $\|av\| = |a| \, \|v\|$ for all $a \in \mathbb{F}$ and $v \in V$;
(3) **Triangle inequality:** $\|v + w\| \leq \|v\| + \|w\|$ for all $v, w \in V$.

Remark 9.2.2. Note that, in fact, $\|v\| \geq 0$ for each $v \in V$ since
$$0 = \|v - v\| \leq \|v\| + \| - v\| = 2\|v\|.$$

Next we want to show that a norm can always be defined from an inner product $\langle \cdot, \cdot \rangle$ via the formula
$$\|v\| = \sqrt{\langle v, v \rangle} \quad \text{for all } v \in V. \tag{9.1}$$
Properties (1) and (2) follow easily from Conditions (1) and (3) of Definition 9.1.1. The triangle inequality requires more careful proof, though, which we give in Theorem 9.3.4 below.

If we take $V = \mathbb{R}^n$, then the norm defined by the usual dot product is related to the usual notion of length of a vector. Namely, for $v = (x_1, \ldots, x_n) \in \mathbb{R}^n$, we have
$$\|v\| = \sqrt{x_1^2 + \cdots + x_n^2}. \tag{9.2}$$
We illustrate this for the case of $\mathbb{R}^3$ in Figure 9.1.

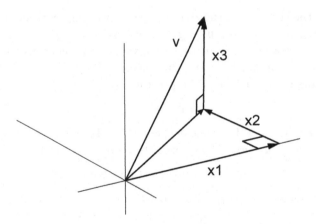

Fig. 9.1 The length of a vector in $\mathbb{R}^3$ via Equation 9.2

While it is always possible to start with an inner product and use it to define a norm, the converse does not hold in general. One can prove that a norm can be written in terms of an inner product as in Equation (9.1) if and only if the norm satisfies the Parallelogram Law (Theorem 9.3.6).

9.3   Orthogonality

Using the inner product, we can now define the notion of orthogonality, prove that the Pythagorean theorem holds in any inner product space, and use the Cauchy-Schwarz inequality to prove the triangle inequality. In particular, this will show that $\|v\| = \sqrt{\langle v, v \rangle}$ does indeed define a norm.

Definition 9.3.1. Two vectors $u, v \in V$ are **orthogonal** (denoted $u \perp v$) if $\langle u, v \rangle = 0$.

Note that the zero vector is the only vector that is orthogonal to itself. In fact, the zero vector is orthogonal to every vector $v \in V$.

Theorem 9.3.2 (Pythagorean Theorem). *If $u, v \in V$, an inner product space, with $u \perp v$, then $\| \cdot \|$ defined by $\|v\| := \sqrt{\langle v, v \rangle}$ obeys*

$$\|u + v\|^2 = \|u\|^2 + \|v\|^2.$$

Proof. Suppose $u, v \in V$ such that $u \perp v$. Then

$$\|u + v\|^2 = \langle u + v, u + v \rangle = \|u\|^2 + \|v\|^2 + \langle u, v \rangle + \langle v, u \rangle$$
$$= \|u\|^2 + \|v\|^2.$$

$\square$

Note that the converse of the Pythagorean Theorem holds for real vector spaces since, in that case, $\langle u, v \rangle + \langle v, u \rangle = 2\mathrm{Re}\langle u, v \rangle = 0$.

Given two vectors $u, v \in V$ with $v \neq 0$, we can uniquely decompose u into two pieces: one piece parallel to v and one piece orthogonal to v. This is called an **orthogonal decomposition**. More precisely, we have

$$u = u_1 + u_2,$$

where $u_1 = av$ and $u_2 \perp v$ for some scalar $a \in \mathbb{F}$. To obtain such a decomposition, write $u_2 = u - u_1 = u - av$. Then, for u_2 to be orthogonal to v, we need

$$0 = \langle u - av, v \rangle = \langle u, v \rangle - a\|v\|^2.$$

Solving for a yields $a = \langle u, v \rangle / \|v\|^2$ so that

$$u = \frac{\langle u, v \rangle}{\|v\|^2}v + \left(u - \frac{\langle u, v \rangle}{\|v\|^2}v \right). \tag{9.3}$$

This decomposition is particularly useful since it allows us to provide a simple proof for the Cauchy-Schwarz inequality.

Theorem 9.3.3 (Cauchy-Schwarz Inequality). *Given any $u, v \in V$, we have*

$$|\langle u, v \rangle| \leq \|u\|\|v\|.$$

Furthermore, equality holds if and only if u and v are linearly dependent, i.e., are scalar multiples of each other.

Proof. If $v = 0$, then both sides of the inequality are zero. Hence, assume that $v \neq 0$, and consider the orthogonal decomposition

$$u = \frac{\langle u, v \rangle}{\|v\|^2} v + w$$

where $w \perp v$. By the Pythagorean theorem, we have

$$\|u\|^2 = \left\| \frac{\langle u, v \rangle}{\|v\|^2} v \right\|^2 + \|w\|^2 = \frac{|\langle u, v \rangle|^2}{\|v\|^2} + \|w\|^2 \geq \frac{|\langle u, v \rangle|^2}{\|v\|^2}.$$

Multiplying both sides by $\|v\|^2$ and taking the square root then yields the Cauchy-Schwarz inequality.

Note that we get equality in the above arguments if and only if $w = 0$. But, by Equation (9.3), this means that u and v are linearly dependent. $\square$

The Cauchy-Schwarz inequality has many different proofs. Here is another one.

Alternate proof of Theorem 9.3.3. Given $u, v \in V$, consider the norm square of the vector $u + re^{i\theta}v$:

$$0 \leq \|u + re^{i\theta}v\|^2 = \|u\|^2 + r^2\|v\|^2 + 2\mathrm{Re}(re^{i\theta}\langle u, v \rangle).$$

Since $\langle u, v \rangle$ is a complex number, one can choose θ so that $e^{i\theta}\langle u, v \rangle$ is real. Hence, the right-hand side is a parabola $ar^2 + br + c$ with real coefficients. It will lie above the real axis, i.e., $ar^2 + br + c \geq 0$, if it does not have any real solutions for r. This is the case when the discriminant satisfies $b^2 - 4ac \leq 0$. In our case this means

$$4|\langle u, v \rangle|^2 - 4\|u\|^2\|v\|^2 \leq 0.$$

Moreover, equality only holds if r can be chosen such that $u + re^{i\theta}v = 0$, which means that u and v are scalar multiples. $\square$

Now that we have proven the Cauchy-Schwarz inequality, we are finally able to verify the triangle inequality. This is the final step in showing that $\|v\| = \sqrt{\langle v, v \rangle}$ does indeed define a norm. We illustrate the triangle inequality in Figure 9.2.

Theorem 9.3.4 (Triangle Inequality). *For all $u, v \in V$ we have*

$$\|u + v\| \leq \|u\| + \|v\|.$$

Proof. By a straightforward calculation, we obtain

$$\|u + v\|^2 = \langle u + v, u + v \rangle = \langle u, u \rangle + \langle v, v \rangle + \langle u, v \rangle + \langle v, u \rangle$$

$$= \langle u, u \rangle + \langle v, v \rangle + \langle u, v \rangle + \overline{\langle u, v \rangle} = \|u\|^2 + \|v\|^2 + 2\mathrm{Re}\langle u, v \rangle.$$

Note that $\mathrm{Re}\langle u, v \rangle \leq |\langle u, v \rangle|$ so that, using the Cauchy-Schwarz inequality, we obtain

$$\|u + v\|^2 \leq \|u\|^2 + \|v\|^2 + 2\|u\|\|v\| = (\|u\| + \|v\|)^2.$$

Taking the square root of both sides now gives the triangle inequality. $\square$

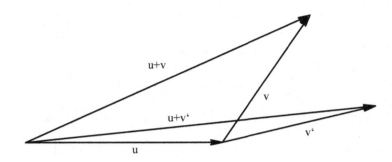

Fig. 9.2   The triangle inequality in $\mathbb{R}^2$

Remark 9.3.5. Note that equality holds for the triangle inequality if and only if $v = ru$ or $u = rv$ for some $r \geq 0$. Namely, equality in the proof happens only if $\langle u, v \rangle = \|u\|\|v\|$, which is equivalent to u and v being scalar multiples of one another.

Theorem 9.3.6 (Parallelogram Law). *Given any* $u, v \in V$, *we have*

$$\|u + v\|^2 + \|u - v\|^2 = 2(\|u\|^2 + \|v\|^2).$$

Proof. By direct calculation,

$$
\begin{aligned}
\|u + v\|^2 + \|u - v\|^2 &= \langle u + v, u + v \rangle + \langle u - v, u - v \rangle \\
&= \|u\|^2 + \|v\|^2 + \langle u, v \rangle + \langle v, u \rangle + \|u\|^2 + \|v\|^2 - \langle u, v \rangle - \langle v, u \rangle \\
&= 2(\|u\|^2 + \|v\|^2).
\end{aligned}
$$

$\square$

Remark 9.3.7. We illustrate the parallelogram law in Figure 9.3.

9.4   Orthonormal bases

We now define the notions of orthogonal basis and orthonormal basis for an inner product space. As we will see later, orthonormal bases have special properties that lead to useful simplifications in common linear algebra calculations.

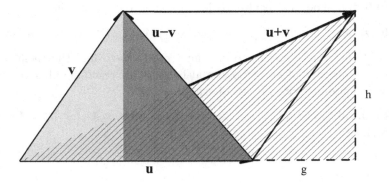

Fig. 9.3   The parallelogram law in $\mathbb{R}^2$

Definition 9.4.1. Let V be an inner product space with inner product $\langle \cdot, \cdot \rangle$. A list of non-zero vectors $(e_1, \ldots, e_m)$ in V is called **orthogonal** if

$$\langle e_i, e_j \rangle = 0, \quad \text{for all } 1 \le i \ne j \le m.$$

The list $(e_1, \ldots, e_m)$ is called **orthonormal** if

$$\langle e_i, e_j \rangle = \delta_{i,j}, \quad \text{for all } i, j = 1, \ldots, m,$$

where δ_{ij} is the Kronecker delta symbol. I.e., $\delta_{ij} = 1$ if $i = j$ and is zero otherwise.

Proposition 9.4.2. *Every orthogonal list of non-zero vectors in V is linearly independent.*

Proof. Let $(e_1, \ldots, e_m)$ be an orthogonal list of vectors in V, and suppose that $a_1, \ldots, a_m \in \mathbb{F}$ are such that

$$a_1 e_1 + \cdots + a_m e_m = 0.$$

Then

$$0 = \|a_1 e_1 + \cdots + a_m e_m\|^2 = |a_1|^2 \|e_1\|^2 + \cdots + |a_m|^2 \|e_m\|^2.$$

Note that $\|e_k\| > 0$, for all $k = 1, \ldots, m$, since every e_k is a non-zero vector. Also, $|a_k|^2 \ge 0$. Hence, the only solution to $a_1 e_1 + \cdots + a_m e_m = 0$ is $a_1 = \cdots = a_m = 0$. $\qquad \square$

Definition 9.4.3. An **orthonormal basis** of a finite-dimensional inner product space V is a list of orthonormal vectors that is basis for V.

Clearly, any orthonormal list of length $\dim(V)$ is an orthonormal basis for V.

Example 9.4.4. The canonical basis for $\mathbb{F}^n$ is an orthonormal basis.

Example 9.4.5. The list $((\frac{1}{\sqrt{2}}, \frac{1}{\sqrt{2}}), (\frac{1}{\sqrt{2}}, -\frac{1}{\sqrt{2}}))$ is an orthonormal basis for $\mathbb{R}^2$.

The next theorem allows us to use inner products to find the coefficients of a vector $v \in V$ in terms of an orthonormal basis. This result highlights how much easier it is to compute with an orthonormal basis.

Theorem 9.4.6. *Let $(e_1, \ldots, e_n)$ be an orthonormal basis for V. Then, for all $v \in V$, we have*

$$v = \langle v, e_1 \rangle e_1 + \cdots + \langle v, e_n \rangle e_n$$

and $\|v\|^2 = \sum_{k=1}^{n} |\langle v, e_k \rangle|^2$.

Proof. Let $v \in V$. Since $(e_1, \ldots, e_n)$ is a basis for V, there exist unique scalars $a_1, \ldots, a_n \in \mathbb{F}$ such that

$$v = a_1 e_1 + \cdots + a_n e_n.$$

Taking the inner product of both sides with respect to e_k then yields $\langle v, e_k \rangle = a_k$. $\square$

9.5   The Gram-Schmidt orthogonalization procedure

We now come to a fundamentally important algorithm, which is called the **Gram-Schmidt orthogonalization procedure**. This algorithm makes it possible to construct, for each list of linearly independent vectors (resp. basis) in an inner product space, a corresponding orthonormal list (resp. orthonormal basis).

Theorem 9.5.1. *If $(v_1, \ldots, v_m)$ is a list of linearly independent vectors in an inner product space V, then there exists an orthonormal list $(e_1, \ldots, e_m)$ such that*

$$\mathrm{span}(v_1, \ldots, v_k) = \mathrm{span}(e_1, \ldots, e_k), \quad \text{for all } k = 1, \ldots, m. \tag{9.4}$$

Proof. The proof is constructive, that is, we will actually construct vectors $e_1, \ldots, e_m$ having the desired properties. Since $(v_1, \ldots, v_m)$ is linearly independent, $v_k \neq 0$ for each $k = 1, 2, \ldots, m$. Set $e_1 = \frac{v_1}{\|v_1\|}$. Then e_1 is a vector of norm 1 and satisfies Equation (9.4) for $k = 1$. Next, set

$$e_2 = \frac{v_2 - \langle v_2, e_1 \rangle e_1}{\|v_2 - \langle v_2, e_1 \rangle e_1\|}.$$

This is, in fact, the normalized version of the orthogonal decomposition Equation (9.3). I.e.,

$$w = v_2 - \langle v_2, e_1 \rangle e_1,$$

where $w \perp e_1$. Note that $\|e_2\| = 1$ and $\mathrm{span}(e_1, e_2) = \mathrm{span}(v_1, v_2)$.

Now, suppose that $e_1, \ldots, e_{k-1}$ have been constructed such that $(e_1, \ldots, e_{k-1})$ is an orthonormal list and $\mathrm{span}(v_1, \ldots, v_{k-1}) = \mathrm{span}(e_1, \ldots, e_{k-1})$. Then define

$$e_k = \frac{v_k - \langle v_k, e_1 \rangle e_1 - \langle v_k, e_2 \rangle e_2 - \cdots - \langle v_k, e_{k-1} \rangle e_{k-1}}{\| v_k - \langle v_k, e_1 \rangle e_1 - \langle v_k, e_2 \rangle e_2 - \cdots - \langle v_k, e_{k-1} \rangle e_{k-1} \|}.$$

Since $(v_1, \ldots, v_k)$ is linearly independent, we know that $v_k \notin \mathrm{span}(v_1, \ldots, v_{k-1})$. Hence, we also know that $v_k \notin \mathrm{span}(e_1, \ldots, e_{k-1})$. It follows that the norm in the definition of e_k is not zero, and so e_k is well-defined (i.e., we are not dividing by zero). Note that a vector divided by its norm has norm 1 so that $\| e_k \| = 1$. Furthermore,

$$\langle e_k, e_i \rangle = \left\langle \frac{v_k - \langle v_k, e_1 \rangle e_1 - \langle v_k, e_2 \rangle e_2 - \cdots - \langle v_k, e_{k-1} \rangle e_{k-1}}{\| v_k - \langle v_k, e_1 \rangle e_1 - \langle v_k, e_2 \rangle e_2 - \cdots - \langle v_k, e_{k-1} \rangle e_{k-1} \|}, e_i \right\rangle$$

$$= \frac{\langle v_k, e_i \rangle - \langle v_k, e_i \rangle}{\| v_k - \langle v_k, e_1 \rangle e_1 - \langle v_k, e_2 \rangle e_2 - \cdots - \langle v_k, e_{k-1} \rangle e_{k-1} \|} = 0,$$

for each $1 \leq i < k$. Hence, $(e_1, \ldots, e_k)$ is orthonormal.

From the definition of e_k, we see that $v_k \in \mathrm{span}(e_1, \ldots, e_k)$ so that $\mathrm{span}(v_1, \ldots, v_k) \subset \mathrm{span}(e_1, \ldots, e_k)$. Since both lists $(e_1, \ldots, e_k)$ and $(v_1, \ldots, v_k)$ are linearly independent, they must span subspaces of the same dimension and therefore are the same subspace. Hence Equation (9.4) holds. $\square$

Example 9.5.2. Take $v_1 = (1, 1, 0)$ and $v_2 = (2, 1, 1)$ in $\mathbb{R}^3$. The list (v_1, v_2) is linearly independent (as you should verify!). To illustrate the Gram-Schmidt procedure, we begin by setting

$$e_1 = \frac{v_1}{\| v_1 \|} = \frac{1}{\sqrt{2}}(1, 1, 0).$$

Next, set

$$e_2 = \frac{v_2 - \langle v_2, e_1 \rangle e_1}{\| v_2 - \langle v_2, e_1 \rangle e_1 \|}.$$

The inner product $\langle v_2, e_1 \rangle = \frac{1}{\sqrt{2}} \langle (1, 1, 0), (2, 1, 1) \rangle = \frac{3}{\sqrt{2}}$, so

$$u_2 = v_2 - \langle v_2, e_1 \rangle e_1 = (2, 1, 1) - \frac{3}{2}(1, 1, 0) = \frac{1}{2}(1, -1, 2).$$

Calculating the norm of u_2, we obtain $\| u_2 \| = \sqrt{\frac{1}{4}(1 + 1 + 4)} = \frac{\sqrt{6}}{2}$. Hence, normalizing this vector, we obtain

$$e_2 = \frac{u_2}{\| u_2 \|} = \frac{1}{\sqrt{6}}(1, -1, 2).$$

The list (e_1, e_2) is therefore orthonormal and has the same span as (v_1, v_2).

Corollary 9.5.3. *Every finite-dimensional inner product space has an orthonormal basis.*

Proof. Let $(v_1, \ldots, v_n)$ be any basis for V. This list is linearly independent and spans V. Apply the Gram-Schmidt procedure to this list to obtain an orthonormal list $(e_1, \ldots, e_n)$, which still spans V by construction. By Proposition 9.4.2, this list is linearly independent and hence a basis of V. $\square$

Corollary 9.5.4. *Every orthonormal list of vectors in V can be extended to an orthonormal basis of V.*

Proof. Let $(e_1, \ldots, e_m)$ be an orthonormal list of vectors in V. By Proposition 9.4.2, this list is linearly independent and hence can be extended to a basis $(e_1, \ldots, e_m, v_1, \ldots, v_k)$ of V by the Basis Extension Theorem. Now apply the Gram-Schmidt procedure to obtain a new orthonormal basis $(e_1, \ldots, e_m, f_1, \ldots, f_k)$. The first m vectors do not change since they are already orthonormal. The list still spans V and is linearly independent by Proposition 9.4.2 and therefore forms a basis. $\square$

Recall Theorem 7.5.3: given an operator $T \in \mathcal{L}(V, V)$ on a complex vector space V, there exists a basis B for V such that the matrix $M(T)$ of T with respect to B is upper triangular. We would like to extend this result to require the additional property of orthonormality.

Corollary 9.5.5. *Let V be an inner product space over $\mathbb{F}$ and $T \in \mathcal{L}(V, V)$. If T is upper triangular with respect to some basis, then T is upper triangular with respect to some orthonormal basis.*

Proof. Let $(v_1, \ldots, v_n)$ be a basis of V with respect to which T is upper triangular. Apply the Gram-Schmidt procedure to obtain an orthonormal basis $(e_1, \ldots, e_n)$, and note that

$$\text{span}(e_1, \ldots, e_k) = \text{span}(v_1, \ldots, v_k), \quad \text{for all } 1 \le k \le n.$$

We proved before that T is upper triangular with respect to a basis $(v_1, \ldots, v_n)$ if and only if $\text{span}(v_1, \ldots, v_k)$ is invariant under T for each $1 \le k \le n$. Since these spans are unchanged by the Gram-Schmidt procedure, T is still upper triangular for the corresponding orthonormal basis. $\square$

9.6 Orthogonal projections and minimization problems

Definition 9.6.1. Let V be a finite-dimensional inner product space and $U \subset V$ be a subset (but not necessarily a subspace) of V. Then the **orthogonal complement** of U is defined to be the set

$$U^\perp = \{v \in V \mid \langle u, v \rangle = 0 \text{ for all } u \in U\}.$$

Note that, in fact, $U^\perp$ is always a subspace of V (as you should check!) and that

$$\{0\}^\perp = V \quad \text{and} \quad V^\perp = \{0\}.$$

In addition, if U_1 and U_2 are subsets of V satisfying $U_1 \subset U_2$, then $U_2^\perp \subset U_1^\perp$.

Remarkably, if $U \subset V$ is a subspace of V, then we can say quite a bit more about $U^\perp$.

Theorem 9.6.2. *If $U \subset V$ is a subspace of V, then $V = U \oplus U^\perp$.*

Proof. We need to show two things:

(1) $V = U + U^\perp$.
(2) $U \cap U^\perp = \{0\}$.

To show that Condition (1) holds, let $(e_1, \ldots, e_m)$ be an orthonormal basis of U. Then, for all $v \in V$, we can write

$$v = \underbrace{\langle v, e_1 \rangle e_1 + \cdots + \langle v, e_m \rangle e_m}_{u} + \underbrace{v - \langle v, e_1 \rangle e_1 - \cdots - \langle v, e_m \rangle e_m}_{w}. \tag{9.5}$$

The vector $u \in U$, and

$$\langle w, e_j \rangle = \langle v, e_j \rangle - \langle v, e_j \rangle = 0, \qquad \text{for all } j = 1, 2, \ldots, m,$$

since $(e_1, \ldots, e_m)$ is an orthonormal list of vectors. Hence, $w \in U^\perp$. This implies that $V = U + U^\perp$.

To prove that Condition (2) also holds, let $v \in U \cap U^\perp$. Then v has to be orthogonal to every vector in U, including to itself, and so $\langle v, v \rangle = 0$. However, this implies $v = 0$ so that $U \cap U^\perp = \{0\}$. $\qquad\square$

Example 9.6.3. $\mathbb{R}^2$ is the direct sum of any two orthogonal lines, and $\mathbb{R}^3$ is the direct sum of any plane and any line orthogonal to the plane (as illustrated in Figure 9.4). For example,

$$\mathbb{R}^2 = \{(x, 0) \mid x \in \mathbb{R}\} \oplus \{(0, y) \mid y \in \mathbb{R}\},$$
$$\mathbb{R}^3 = \{(x, y, 0) \mid x, y \in \mathbb{R}\} \oplus \{(0, 0, z) \mid z \in \mathbb{R}\}.$$

Another fundamental fact about the orthogonal complement of a subspace is as follows.

Theorem 9.6.4. *If $U \subset V$ is a subspace of V, then $U = (U^\perp)^\perp$.*

Proof. First we show that $U \subset (U^\perp)^\perp$. Let $u \in U$. Then, for all $v \in U^\perp$, we have $\langle u, v \rangle = 0$. Hence, $u \in (U^\perp)^\perp$ by the definition of $(U^\perp)^\perp$.

Next we show that $(U^\perp)^\perp \subset U$. Suppose $0 \neq v \in (U^\perp)^\perp$ such that $v \notin U$, and decompose v according to Theorem 9.6.2, i.e., as

$$v = u_1 + u_2 \in U \oplus U^\perp$$

with $u_1 \in U$ and $u_2 \in U^\perp$. Then $u_2 \neq 0$ since $v \notin U$. Furthermore, $\langle u_2, v \rangle = \langle u_2, u_2 \rangle \neq 0$. But then v is not in $(U^\perp)^\perp$, which contradicts our initial assumption. Hence, we must have that $(U^\perp)^\perp \subset U$. $\qquad\square$

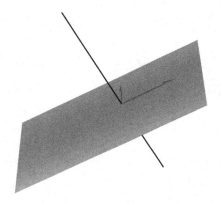

Fig. 9.4 $\mathbb{R}^3$ as a direct sum of a plane and a line, as in Example 9.6.3

By Theorem 9.6.2, we have the decomposition $V = U \oplus U^\perp$ for every subspace $U \subset V$. This allows us to define the **orthogonal projection** P_U of V onto U.

Definition 9.6.5. Let $U \subset V$ be a subspace of a finite-dimensional inner product space. Every $v \in V$ can be uniquely written as $v = u + w$ where $u \in U$ and $w \in U^\perp$. Define

$$P_U : V \to V,$$
$$v \mapsto u.$$

Note that P_U is called a projection operator since it satisfies $P_U^2 = P_U$. Further, since we also have

$$\text{range}\,(P_U) = U,$$
$$\text{null}\,(P_U) = U^\perp,$$

it follows that $\text{range}\,(P_U) \perp \text{null}\,(P_U)$. Therefore, P_U is called an orthogonal projection.

The decomposition of a vector $v \in V$ as given in Equation (9.5) yields the formula

$$P_U v = \langle v, e_1 \rangle e_1 + \cdots + \langle v, e_m \rangle e_m, \qquad (9.6)$$

where $(e_1, \dots, e_m)$ is any orthonormal basis of U. Equation (9.6) is a particularly useful tool for computing such things as the matrix of P_U with respect to the basis $(e_1, \dots, e_m)$.

Let us now apply the inner product to the following minimization problem: Given a subspace $U \subset V$ and a vector $v \in V$, find the vector $u \in U$ that is closest to the vector v. In other words, we want to make $\|v - u\|$ as small as possible. The next proposition shows that $P_U v$ is the closest point in U to the vector v and that this minimum is, in fact, unique.

Proposition 9.6.6. *Let $U \subset V$ be a subspace of V and $v \in V$. Then*

$$\|v - P_U v\| \le \|v - u\| \qquad \text{for every } u \in U.$$

Furthermore, equality holds if and only if $u = P_U v$.

Proof. Let $u \in U$ and set $P := P_U$ for short. Then

$$\|v - Pv\|^2 \le \|v - Pv\|^2 + \|Pv - u\|^2$$
$$= \|(v - Pv) + (Pv - u)\|^2 = \|v - u\|^2,$$

where the second line follows from the Pythagorean Theorem 9.3.2 since $v - Pv \in U^\perp$ and $Pv - u \in U$. Furthermore, equality holds only if $\|Pv - u\|^2 = 0$, which is equivalent to $Pv = u$. $\qquad\qquad \square$

Example 9.6.7. Consider the plane $U \subset \mathbb{R}^3$ through 0 and perpendicular to the vector $u = (1, 1, 1)$. Using the standard norm on $\mathbb{R}^3$, we can calculate the distance of the point $v = (1, 2, 3)$ to U using Proposition 9.6.6. In particular, the distance d between v and U is given by $d = \|v - P_U v\|$. Let $(\frac{1}{\sqrt{3}}u, u_1, u_2)$ be a basis for $\mathbb{R}^3$ such that (u_1, u_2) is an orthonormal basis of U. Then, by Equation (9.6), we have

$$v - P_U v = (\frac{1}{3}\langle v, u\rangle u + \langle v, u_1\rangle u_1 + \langle v, u_2\rangle u_2) - (\langle v, u_1\rangle u_1 + \langle v, u_2\rangle u_2)$$
$$= \frac{1}{3}\langle v, u\rangle u$$
$$= \frac{1}{3}\langle (1, 2, 3), (1, 1, 1)\rangle (1, 1, 1)$$
$$= (2, 2, 2).$$

Hence, $d = \|(2, 2, 2)\| = 2\sqrt{3}$.

Exercises for Chapter 9

Calculational Exercises

(1) Let (e_1, e_2, e_3) be the canonical basis of $\mathbb{R}^3$, and define

$$f_1 = e_1 + e_2 + e_3$$
$$f_2 = e_2 + e_3$$
$$f_3 = e_3.$$

   (a) Apply the Gram-Schmidt process to the basis (f_1, f_2, f_3).

   (b) What do you obtain if you instead applied the Gram-Schmidt process to the basis (f_3, f_2, f_1)?

(2) Let $\mathcal{C}[-\pi, \pi] = \{f : [-\pi, \pi] \to \mathbb{R} \mid f \text{ is continuous}\}$ denote the inner product space of continuous real-valued functions defined on the interval $[-\pi, \pi] \subset \mathbb{R}$, with inner product given by

$$\langle f, g \rangle = \int_{-\pi}^{\pi} f(x)g(x)dx, \text{ for every } f, g \in \mathcal{C}[-\pi, \pi].$$

Then, given any positive integer $n \in \mathbb{Z}_+$, verify that the set of vectors

$$\left\{ \frac{1}{\sqrt{2\pi}}, \frac{\sin(x)}{\sqrt{\pi}}, \frac{\sin(2x)}{\sqrt{\pi}}, \dots, \frac{\sin(nx)}{\sqrt{\pi}}, \frac{\cos(x)}{\sqrt{\pi}}, \frac{\cos(2x)}{\sqrt{\pi}}, \dots, \frac{\cos(nx)}{\sqrt{\pi}} \right\}$$

is orthonormal.

(3) Let $\mathbb{R}_2[x]$ denote the inner product space of polynomials over $\mathbb{R}$ having degree at most two, with inner product given by

$$\langle f, g \rangle = \int_0^1 f(x)g(x)dx, \text{ for every } f, g \in \mathbb{R}_2[x].$$

Apply the Gram-Schmidt procedure to the standard basis $\{1, x, x^2\}$ for $\mathbb{R}_2[x]$ in order to produce an orthonormal basis for $\mathbb{R}_2[x]$.

(4) Let $v_1, v_2, v_3 \in \mathbb{R}^3$ be given by $v_1 = (1, 2, 1)$, $v_2 = (1, -2, 1)$, and $v_3 = (1, 2, -1)$. Apply the Gram-Schmidt procedure to the basis (v_1, v_2, v_3) of $\mathbb{R}^3$, and call the resulting orthonormal basis (u_1, u_2, u_3).

(5) Let $P \subset \mathbb{R}^3$ be the plane containing 0 perpendicular to the vector $(1, 1, 1)$. Using the standard norm, calculate the distance of the point $(1, 2, 3)$ to P.

(6) Give an orthonormal basis for $\text{null}(T)$, where $T \in \mathcal{L}(\mathbb{C}^4)$ is the map with canonical matrix

$$\begin{pmatrix} 1 & 1 & 1 & 1 \\ 1 & 1 & 1 & 1 \\ 1 & 1 & 1 & 1 \\ 1 & 1 & 1 & 1 \end{pmatrix}.$$

Proof-Writing Exercises

(1) Let V be a finite-dimensional inner product space over $\mathbb{F}$. Given any vectors $u, v \in V$, prove that the following two statements are equivalent:

(a) $\langle u, v \rangle = 0$
(b) $\|u\| \leq \|u + \alpha v\|$ for every $\alpha \in \mathbb{F}$.

(2) Let $n \in \mathbb{Z}_+$ be a positive integer, and let $a_1, \dots, a_n, b_1, \dots, b_n \in \mathbb{R}$ be any collection of $2n$ real numbers. Prove that

$$\left(\sum_{k=1}^{n} a_k b_k \right)^2 \leq \left(\sum_{k=1}^{n} k a_k^2 \right) \left(\sum_{k=1}^{n} \frac{b_k^2}{k} \right).$$

(3) Prove or disprove the following claim:

Claim. *There is an inner product $\langle \cdot, \cdot \rangle$ on $\mathbb{R}^2$ whose associated norm $\| \cdot \|$ is given by the formula*

$$\|(x_1, x_2)\| = |x_1| + |x_2|$$

for every vector $(x_1, x_2) \in \mathbb{R}^2$, where $|\cdot|$ denotes the absolute value function on $\mathbb{R}$.

(4) Let V be a finite-dimensional inner product space over $\mathbb{R}$. Given $u, v \in V$, prove that

$$\langle u, v \rangle = \frac{\|u+v\|^2 - \|u-v\|^2}{4}.$$

(5) Let V be a finite-dimensional inner product space over $\mathbb{C}$. Given $u, v \in V$, prove that

$$\langle u, v \rangle = \frac{\|u+v\|^2 - \|u-v\|^2}{4} + \frac{\|u+iv\|^2 - \|u-iv\|^2}{4}i.$$

(6) Let V be a finite-dimensional inner product space over $\mathbb{F}$, and let U be a subspace of V. Prove that the orthogonal complement $U^\perp$ of U with respect to the inner product $\langle \cdot, \cdot \rangle$ on V satisfies

$$\dim(U^\perp) = \dim(V) - \dim(U).$$

(7) Let V be a finite-dimensional inner product space over $\mathbb{F}$, and let U be a subspace of V. Prove that $U = V$ if and only if the orthogonal complement $U^\perp$ of U with respect to the inner product $\langle \cdot, \cdot \rangle$ on V satisfies $U^\perp = \{0\}$.

(8) Let V be a finite-dimensional inner product space over $\mathbb{F}$, and suppose that $P \in \mathcal{L}(V)$ is a linear operator on V having the following two properties:

(a) Given any vector $v \in V$, $P(P(v)) = P(v)$. I.e., $P^2 = P$.
(b) Given any vector $u \in \text{null}(P)$ and any vector $v \in \text{range}(P)$, $\langle u, v \rangle = 0$.

Prove that P is an orthogonal projection.

(9) Prove or give a counterexample: For any $n \geq 1$ and $A \in \mathbb{C}^{n \times n}$, one has

$$\text{null}(A) = (\text{range}(A))^\perp.$$

(10) Prove or give a counterexample: The Gram-Schmidt process applied to an orthonormal list of vectors reproduces that list unchanged.

Chapter 10

Change of Bases

In Section 6.6, we saw that linear operators on an n-dimensional vector space are in one-to-one correspondence with $n \times n$ matrices. This correspondence, however, depends upon the choice of basis for the vector space. In this chapter we address the question of how the matrix for a linear operator changes if we change from one orthonormal basis to another.

10.1   Coordinate vectors

Let V be a finite-dimensional inner product space with inner product $\langle \cdot, \cdot \rangle$ and dimension $\dim(V) = n$. Then V has an orthonormal basis $e = (e_1, \ldots, e_n)$, and, according to Theorem 9.4.6, every $v \in V$ can be written as

$$v = \sum_{i=1}^{n} \langle v, e_i \rangle e_i.$$

This induces a map

$$[\,\cdot\,]_e : V \to \mathbb{F}^n$$

$$v \mapsto \begin{bmatrix} \langle v, e_1 \rangle \\ \vdots \\ \langle v, e_n \rangle \end{bmatrix},$$

which maps the vector $v \in V$ to the $n \times 1$ column vector of its coordinates with respect to the basis e. The column vector $[v]_e$ is called the **coordinate vector** of v with respect to the basis e.

Example 10.1.1. Recall that the vector space $\mathbb{R}_1[x]$ of polynomials over $\mathbb{R}$ of degree at most 1 is an inner product space with inner product defined by

$$\langle f, g \rangle = \int_0^1 f(x)g(x)dx.$$

Then $e = (1, \sqrt{3}(-1 + 2x))$ forms an orthonormal basis for $\mathbb{R}_1[x]$. The coordinate vector of the polynomial $p(x) = 3x + 2 \in \mathbb{R}_1[x]$ is, e.g.,

$$[p(x)]_e = \frac{1}{2}\begin{bmatrix} 7 \\ \sqrt{3} \end{bmatrix}.$$

111

Note also that the map $[\,\cdot\,]_e$ is an isomorphism (meaning that it is an injective and surjective linear map) and that it is also inner product preserving. Denote the usual inner product on $\mathbb{F}^n$ by

$$\langle x, y \rangle_{\mathbb{F}^n} = \sum_{k=1}^{n} x_k \overline{y}_k.$$

Then

$$\langle v, w \rangle_V = \langle [v]_e, [w]_e \rangle_{\mathbb{F}^n}, \qquad \text{for all } v, w \in V,$$

since

$$\langle v, w \rangle_V = \sum_{i,j=1}^{n} \langle \langle v, e_i \rangle e_i, \langle w, e_j \rangle e_j \rangle = \sum_{i,j=1}^{n} \langle v, e_i \rangle \overline{\langle w, e_j \rangle} \langle e_i, e_j \rangle$$

$$= \sum_{i,j=1}^{n} \langle v, e_i \rangle \overline{\langle w, e_j \rangle} \delta_{ij} = \sum_{i=1}^{n} \langle v, e_i \rangle \overline{\langle w, e_i \rangle} = \langle [v]_e, [w]_e \rangle_{\mathbb{F}^n}.$$

It is important to remember that the map $[\,\cdot\,]_e$ depends on the choice of basis $e = (e_1, \ldots, e_n)$.

10.2   Change of basis transformation

Recall that we can associate a matrix $A \in \mathbb{F}^{n \times n}$ to every operator $T \in \mathcal{L}(V, V)$. More precisely, the jth column of the matrix $A = M(T)$ with respect to a basis $e = (e_1, \ldots, e_n)$ is obtained by expanding Te_j in terms of the basis e. If the basis e is orthonormal, then the coefficient of e_i is just the inner product of the vector with e_i. Hence,

$$M(T) = (\langle Te_j, e_i \rangle)_{1 \le i, j \le n},$$

where i is the row index and j is the column index of the matrix.

Conversely, if $A \in \mathbb{F}^{n \times n}$ is a matrix, then we can associate a linear operator $T \in \mathcal{L}(V, V)$ to A by setting

$$Tv = \sum_{j=1}^{n} \langle v, e_j \rangle Te_j = \sum_{j=1}^{n} \sum_{i=1}^{n} \langle Te_j, e_i \rangle \langle v, e_j \rangle e_i$$

$$= \sum_{i=1}^{n} \left(\sum_{j=1}^{n} a_{ij} \langle v, e_j \rangle \right) e_i = \sum_{i=1}^{n} (A[v]_e)_i e_i,$$

where $(A[v]_e)_i$ denotes the ith component of the column vector $A[v]_e$. With this construction, we have $M(T) = A$. The coefficients of Tv in the basis $(e_1, \ldots, e_n)$ are recorded by the column vector obtained by multiplying the $n \times n$ matrix A with the $n \times 1$ column vector $[v]_e$ whose components $([v]_e)_j = \langle v, e_j \rangle$.

Example 10.2.1. Given
$$A = \begin{bmatrix} 1 & -i \\ i & 1 \end{bmatrix},$$
we can define $T \in \mathcal{L}(V,V)$ with respect to the canonical basis as follows:
$$T\begin{bmatrix} z_1 \\ z_2 \end{bmatrix} = \begin{bmatrix} 1 & -i \\ i & 1 \end{bmatrix}\begin{bmatrix} z_1 \\ z_2 \end{bmatrix} = \begin{bmatrix} z_1 - iz_2 \\ iz_1 + z_2 \end{bmatrix}.$$

Suppose that we want to use another orthonormal basis $f = (f_1, \ldots, f_n)$ for V. Then, as before, we have $v = \sum_{i=1}^n \langle v, f_i \rangle f_i$. Comparing this with $v = \sum_{j=1}^n \langle v, e_j \rangle e_j$, we find that
$$v = \sum_{i,j=1}^n \langle \langle v, e_j \rangle e_j, f_i \rangle f_i = \sum_{i=1}^n \left(\sum_{j=1}^n \langle e_j, f_i \rangle \langle v, e_j \rangle \right) f_i.$$
Hence,
$$[v]_f = S[v]_e,$$
where
$$S = (s_{ij})_{i,j=1}^n \qquad \text{with } s_{ij} = \langle e_j, f_i \rangle.$$
The jth column of S is given by the coefficients of the expansion of e_j in terms of the basis $f = (f_1, \ldots, f_n)$. The matrix S describes a linear map in $\mathcal{L}(\mathbb{F}^n)$, which is called the **change of basis transformation**.

We may also interchange the role of bases e and f. In this case, we obtain the matrix $R = (r_{ij})_{i,j=1}^n$, where
$$r_{ij} = \langle f_j, e_i \rangle.$$
Then, by the uniqueness of the expansion in a basis, we obtain
$$[v]_e = R[v]_f$$
so that
$$RS[v]_e = [v]_e, \qquad \text{for all } v \in V.$$
Since this equation is true for all $[v]_e \in \mathbb{F}^n$, it follows that either $RS = I$ or $R = S^{-1}$. In particular, S and R are invertible. We can also check this explicitly by using the properties of orthonormal bases. Namely,
$$(RS)_{ij} = \sum_{k=1}^n r_{ik}s_{kj} = \sum_{k=1}^n \langle f_k, e_i \rangle \langle e_j, f_k \rangle$$
$$= \sum_{k=1}^n \langle e_j, f_k \rangle \overline{\langle e_i, f_k \rangle} = \langle [e_j]_f, [e_i]_f \rangle_{\mathbb{F}^n} = \delta_{ij}.$$

Matrix S (and similarly also R) has the interesting property that its columns are orthonormal to one another. This follows from the fact that the columns are the coordinates of orthonormal vectors with respect to another orthonormal basis. A similar statement holds for the rows of S (and similarly also R). More information about orthogonal matrices can be found in Appendix A.5.1, in particular Definition A.5.3.

Example 10.2.2. Let $V = \mathbb{C}^2$, and choose the orthonormal bases $e = (e_1, e_2)$ and $f = (f_1, f_2)$ with

$$e_1 = \begin{bmatrix} 1 \\ 0 \end{bmatrix}, \qquad\qquad e_2 = \begin{bmatrix} 0 \\ 1 \end{bmatrix},$$

$$f_1 = \frac{1}{\sqrt{2}} \begin{bmatrix} 1 \\ 1 \end{bmatrix}, \qquad\qquad f_2 = \frac{1}{\sqrt{2}} \begin{bmatrix} -1 \\ 1 \end{bmatrix}.$$

Then

$$S = \begin{bmatrix} \langle e_1, f_1 \rangle & \langle e_2, f_1 \rangle \\ \langle e_1, f_2 \rangle & \langle e_2, f_2 \rangle \end{bmatrix} = \frac{1}{\sqrt{2}} \begin{bmatrix} 1 & 1 \\ -1 & 1 \end{bmatrix}$$

and

$$R = \begin{bmatrix} \langle f_1, e_1 \rangle & \langle f_2, e_1 \rangle \\ \langle f_1, e_2 \rangle & \langle f_2, e_2 \rangle \end{bmatrix} = \frac{1}{\sqrt{2}} \begin{bmatrix} 1 & -1 \\ 1 & 1 \end{bmatrix}.$$

One can then check explicitly that indeed

$$RS = \frac{1}{2} \begin{bmatrix} 1 & -1 \\ 1 & 1 \end{bmatrix} \begin{bmatrix} 1 & 1 \\ -1 & 1 \end{bmatrix} = \begin{bmatrix} 1 & 0 \\ 0 & 1 \end{bmatrix} = I.$$

So far we have only discussed how the coordinate vector of a given vector $v \in V$ changes under the change of basis from e to f. The next question we can ask is how the matrix $M(T)$ of an operator $T \in \mathcal{L}(V)$ changes if we change the basis. Let A be the matrix of T with respect to the basis $e = (e_1, \ldots, e_n)$, and let B be the matrix for T with respect to the basis $f = (f_1, \ldots, f_n)$. How do we determine B from A? Note that

$$[Tv]_e = A[v]_e$$

so that

$$[Tv]_f = S[Tv]_e = SA[v]_e = SAR[v]_f = SAS^{-1}[v]_f.$$

This implies that

$$B = SAS^{-1}.$$

Example 10.2.3. Continuing Example 10.2.2, let

$$A = \begin{bmatrix} 1 & 1 \\ 1 & 1 \end{bmatrix}$$

be the matrix of a linear operator with respect to the basis e. Then the matrix B with respect to the basis f is given by

$$B = SAS^{-1} = \frac{1}{2} \begin{bmatrix} 1 & 1 \\ -1 & 1 \end{bmatrix} \begin{bmatrix} 1 & 1 \\ 1 & 1 \end{bmatrix} \begin{bmatrix} 1 & -1 \\ 1 & 1 \end{bmatrix} = \frac{1}{2} \begin{bmatrix} 1 & 1 \\ -1 & 1 \end{bmatrix} \begin{bmatrix} 2 & 0 \\ 2 & 0 \end{bmatrix} = \begin{bmatrix} 2 & 0 \\ 0 & 0 \end{bmatrix}.$$

Exercises for Chapter 10

Calculational Exercises

(1) Consider $\mathbb{R}^3$ with two orthonormal bases: the canonical basis $e = (e_1, e_2, e_3)$ and the basis $f = (f_1, f_2, f_3)$, where

$$f_1 = \frac{1}{\sqrt{3}}(1,1,1), \ f_2 = \frac{1}{\sqrt{6}}(1,-2,1), \ f_3 = \frac{1}{\sqrt{2}}(1,0,-1).$$

Find the matrix, S, of the change of basis transformation such that

$$[v]_f = S[v]_e, \quad \text{for all } v \in \mathbb{R}^3,$$

where $[v]_b$ denotes the column vector of v with respect to the basis b.

(2) Let $v \in \mathbb{C}^4$ be the vector given by $v = (1, i, -1, -i)$. Find the matrix (with respect to the canonical basis on $\mathbb{C}^4$) of the orthogonal projection $P \in \mathcal{L}(\mathbb{C}^4)$ such that

$$\text{null}(P) = \{v\}^\perp.$$

(3) Let U be the subspace of $\mathbb{R}^3$ that coincides with the plane through the origin that is perpendicular to the vector $n = (1,1,1) \in \mathbb{R}^3$.

(a) Find an orthonormal basis for U.
(b) Find the matrix (with respect to the canonical basis on $\mathbb{R}^3$) of the orthogonal projection $P \in \mathcal{L}(\mathbb{R}^3)$ onto U, i.e., such that $\text{range}(P) = U$.

(4) Let $V = \mathbb{C}^4$ with its standard inner product. For $\theta \in \mathbb{R}$, let

$$v_\theta = \begin{pmatrix} 1 \\ e^{i\theta} \\ e^{2i\theta} \\ e^{3i\theta} \end{pmatrix} \in \mathbb{C}^4.$$

Find the canonical matrix of the orthogonal projection onto the subspace $\{v_\theta\}^\perp$.

Proof-Writing Exercises

(1) Let V be a finite-dimensional vector space over $\mathbb{F}$ with dimension $n \in \mathbb{Z}_+$, and suppose that $b = (v_1, v_2, \ldots, v_n)$ is a basis for V. Prove that the coordinate vectors $[v_1]_b, [v_2]_b, \ldots, [v_n]_b$ with respect to b form a basis for $\mathbb{F}^n$.

(2) Let V be a finite-dimensional vector space over $\mathbb{F}$, and suppose that $T \in \mathcal{L}(V)$ is a linear operator having the following property: Given any two bases b and c for V, the matrix $M(T, b)$ for T with respect to b is the same as the matrix $M(T, c)$ for T with respect to c. Prove that there exists a scalar $\alpha \in \mathbb{F}$ such that $T = \alpha I_V$, where I_V denotes the identity map on V.

Chapter 11

The Spectral Theorem for Normal Linear Maps

In this chapter we come back to the question of when a linear operator on an inner product space V is diagonalizable. We first introduce the notion of the adjoint (a.k.a. hermitian conjugate) of an operator, and we then use this to define so-called normal operators. The main result of this chapter is the Spectral Theorem, which states that normal operators are diagonal with respect to an orthonormal basis. We use this to show that normal operators are "unitarily diagonalizable" and generalize this notion to finding the singular-value decomposition of an operator. In this chapter, we will always assume $\mathbb{F} = \mathbb{C}$.

11.1 Self-adjoint or hermitian operators

Let V be a finite-dimensional inner product space over $\mathbb{C}$ with inner product $\langle \cdot, \cdot \rangle$. A linear operator $T \in \mathcal{L}(V)$ is uniquely determined by the values of

$$\langle Tv, w \rangle, \quad \text{for all } v, w \in V.$$

This means, in particular, that if $T, S \in \mathcal{L}(V)$ and

$$\langle Tv, w \rangle = \langle Sv, w \rangle \quad \text{for all } v, w \in V,$$

then $T = S$. To see this, take w to be the elements of an orthonormal basis of V.

Definition 11.1.1. Given $T \in \mathcal{L}(V)$, the **adjoint** (a.k.a. **hermitian conjugate**) of T is defined to be the operator $T^* \in \mathcal{L}(V)$ for which

$$\langle Tv, w \rangle = \langle v, T^*w \rangle, \quad \text{for all } v, w \in V.$$

Moreover, we call T **self-adjoint** (a.k.a. **hermitian**) if $T = T^*$.

The uniqueness of T^* is clear by the previous observation.

Example 11.1.2. Let $V = \mathbb{C}^3$, and let $T \in \mathcal{L}(\mathbb{C}^3)$ be defined by $T(z_1, z_2, z_3) = (2z_2 + iz_3, iz_1, z_2)$. Then

$$
\begin{aligned}
\langle (y_1, y_2, y_3), T^*(z_1, z_2, z_3) \rangle &= \langle T(y_1, y_2, y_3), (z_1, z_2, z_3) \rangle \\
&= \langle (2y_2 + iy_3, iy_1, y_2), (z_1, z_2, z_3) \rangle \\
&= 2y_2\overline{z_1} + iy_3\overline{z_1} + iy_1\overline{z_2} + y_2\overline{z_3} \\
&= \langle (y_1, y_2, y_3), (-iz_2, 2z_1 + z_3, -iz_1) \rangle
\end{aligned}
$$

so that $T^*(z_1, z_2, z_3) = (-iz_2, 2z_1 + z_3, -iz_1)$. Writing the matrix for T in terms of the canonical basis, we see that

$$M(T) = \begin{bmatrix} 0 & 2 & i \\ i & 0 & 0 \\ 0 & 1 & 0 \end{bmatrix} \quad \text{and} \quad M(T^*) = \begin{bmatrix} 0 & -i & 0 \\ 2 & 0 & 1 \\ -i & 0 & 0 \end{bmatrix}.$$

Note that $M(T^*)$ can be obtained from $M(T)$ by taking the complex conjugate of each element and then transposing. This operation is called the **conjugate transpose** of $M(T)$, and we denote it by $(M(T))^*$.

We collect several elementary properties of the adjoint operation into the following proposition. You should provide a proof of these results for your own practice.

Proposition 11.1.3. *Let $S, T \in \mathcal{L}(V)$ and $a \in \mathbb{F}$.*

(1) $(S + T)^* = S^* + T^*$.
(2) $(aT)^* = \overline{a}T^*$.
(3) $(T^*)^* = T$.
(4) $I^* = I$.
(5) $(ST)^* = T^*S^*$.
(6) $M(T^*) = M(T)^*$.

When $n = 1$, note that the conjugate transpose of a 1×1 matrix A is just the complex conjugate of its single entry. Hence, requiring A to be self-adjoint ($A = A^*$) amounts to saying that this sole entry is real. Because of the transpose, though, reality is not the same as self-adjointness when $n > 1$, but the analogy does nonetheless carry over to the eigenvalues of self-adjoint operators.

Proposition 11.1.4. *Every eigenvalue of a self-adjoint operator is real.*

Proof. Suppose $\lambda \in \mathbb{C}$ is an eigenvalue of T and that $0 \neq v \in V$ is a corresponding eigenvector such that $Tv = \lambda v$. Then

$$\lambda\|v\|^2 = \langle \lambda v, v \rangle = \langle Tv, v \rangle = \langle v, T^*v \rangle$$
$$= \langle v, Tv \rangle = \langle v, \lambda v \rangle = \overline{\lambda}\langle v, v \rangle = \overline{\lambda}\|v\|^2.$$

This implies that $\lambda = \overline{\lambda}$. $\square$

Example 11.1.5. The operator $T \in \mathcal{L}(V)$ defined by $T(v) = \begin{bmatrix} 2 & 1 + i \\ 1 - i & 3 \end{bmatrix} v$ is self-adjoint, and it can be checked (e.g., using the characteristic polynomial) that the eigenvalues of T are $\lambda = 1, 4$.

11.2 Normal operators

Normal operators are those that commute with their own adjoint. As we will see, this includes many important examples of operations.

Definition 11.2.1. We call $T \in \mathcal{L}(V)$ **normal** if $TT^* = T^*T$.

Given an arbitrary operator $T \in \mathcal{L}(V)$, we have that $TT^* \neq T^*T$ in general. However, both TT^* and T^*T are self-adjoint, and any self-adjoint operator T is normal. We now give a different characterization for normal operators in terms of norms.

Proposition 11.2.2. *Let V be a complex inner product space, and suppose that $T \in \mathcal{L}(V)$ satisfies*

$$\langle Tv, v \rangle = 0, \quad \text{for all } v \in V.$$

Then $T = 0$.

Proof. You should be able to verify that

$$\langle Tu, w \rangle = \frac{1}{4} \{ \langle T(u+w), u+w \rangle - \langle T(u-w), u-w \rangle$$
$$+ i\langle T(u+iw), u+iw \rangle - i\langle T(u-iw), u-iw \rangle \}.$$

Since each term on the right-hand side is of the form $\langle Tv, v \rangle$, we obtain 0 for each $u, w \in V$. Hence $T = 0$. $\qquad\square$

Proposition 11.2.3. *Let $T \in \mathcal{L}(V)$. Then T is normal if and only if*

$$\|Tv\| = \|T^*v\|, \quad \text{for all } v \in V.$$

Proof. Note that

$$T \text{ is normal} \iff T^*T - TT^* = 0$$
$$\iff \langle (T^*T - TT^*)v, v \rangle = 0, \quad \text{for all } v \in V$$
$$\iff \langle TT^*v, v \rangle = \langle T^*Tv, v \rangle, \quad \text{for all } v \in V$$
$$\iff \|Tv\|^2 = \|T^*v\|^2, \quad \text{for all } v \in V.$$

$\qquad\square$

Corollary 11.2.4. *Let $T \in \mathcal{L}(V)$ be a normal operator.*

(1) $\text{null}\,(T) = \text{null}\,(T^*)$.
(2) If $\lambda \in \mathbb{C}$ is an eigenvalue of T, then $\overline{\lambda}$ is an eigenvalue of T^* with the same eigenvector.
(3) If $\lambda, \mu \in \mathbb{C}$ are distinct eigenvalues of T with associated eigenvectors $v, w \in V$, respectively, then $\langle v, w \rangle = 0$.

Proof. Note that Part 1 follows from Proposition 11.2.3 and the positive definiteness of the norm.

To prove Part 2, first verify that if T is normal, then $T - \lambda I$ is also normal with $(T - \lambda I)^* = T^* - \overline{\lambda} I$. Therefore, by Proposition 11.2.3, we have

$$0 = \|(T - \lambda I)v\| = \|(T - \lambda I)^* v\| = \|(T^* - \overline{\lambda} I)v\|,$$

and so v is an eigenvector of T^* with eigenvalue $\overline{\lambda}$.

Using Part 2, note that

$$(\lambda - \mu)\langle v, w \rangle = \langle \lambda v, w \rangle - \langle v, \overline{\mu} w \rangle = \langle Tv, w \rangle - \langle v, T^* w \rangle = 0.$$

Since $\lambda - \mu \neq 0$ it follows that $\langle v, w \rangle = 0$, proving Part 3. $\qquad\square$

11.3 Normal operators and the spectral decomposition

Recall that an operator $T \in \mathcal{L}(V)$ is diagonalizable if there exists a basis B for V such that B consists entirely of eigenvectors for T. The nicest operators on V are those that are diagonalizable with respect to some orthonormal basis for V. In other words, these are the operators for which we can find an orthonormal basis for V that consists of eigenvectors for T. The Spectral Theorem for finite-dimensional complex inner product spaces states that this can be done precisely for normal operators.

Theorem 11.3.1 (Spectral Theorem). *Let V be a finite-dimensional inner product space over $\mathbb{C}$ and $T \in \mathcal{L}(V)$. Then T is normal if and only if there exists an orthonormal basis for V consisting of eigenvectors for T.*

Proof.
("$\Longrightarrow$") Suppose that T is normal. Combining Theorem 7.5.3 and Corollary 9.5.5, there exists an orthonormal basis $e = (e_1, \ldots, e_n)$ for which the matrix $M(T)$ is upper triangular, i.e.,

$$M(T) = \begin{bmatrix} a_{11} & \cdots & a_{1n} \\ & \ddots & \vdots \\ 0 & & a_{nn} \end{bmatrix}.$$

We will show that $M(T)$ is, in fact, diagonal, which implies that the basis elements $e_1, \ldots, e_n$ are eigenvectors of T.

Since $M(T) = (a_{ij})_{i,j=1}^n$ with $a_{ij} = 0$ for $i > j$, we have $Te_1 = a_{11}e_1$ and $T^* e_1 = \sum_{k=1}^n \overline{a}_{1k} e_k$. Thus, by the Pythagorean Theorem and Proposition 11.2.3,

$$|a_{11}|^2 = \|a_{11}e_1\|^2 = \|Te_1\|^2 = \|T^* e_1\|^2 = \|\sum_{k=1}^n \overline{a}_{1k} e_k\|^2 = \sum_{k=1}^n |a_{1k}|^2,$$

from which it follows that $|a_{12}| = \cdots = |a_{1n}| = 0$. Repeating this argument, $\|Te_j\|^2 = |a_{jj}|^2$ and $\|T^* e_j\|^2 = \sum_{k=j}^n |a_{jk}|^2$ so that $a_{ij} = 0$ for all $2 \leq i < j \leq n$.

Hence, T is diagonal with respect to the basis e, and $e_1, \ldots, e_n$ are eigenvectors of T.

("$\Longleftarrow$") Suppose there exists an orthonormal basis $(e_1, \ldots, e_n)$ for V that consists of eigenvectors for T. Then the matrix $M(T)$ with respect to this basis is diagonal. Moreover, $M(T^*) = M(T)^*$ with respect to this basis must also be a diagonal matrix. It follows that $TT^* = T^*T$ since their corresponding matrices commute:

$$M(TT^*) = M(T)M(T^*) = M(T^*)M(T) = M(T^*T).$$

$\square$

The following corollary is the best possible decomposition of a complex vector space V into subspaces that are invariant under a normal operator T. On each subspace $\mathrm{null}\,(T - \lambda_i I)$, the operator T acts just like multiplication by scalar λ_i. In other words,

$$T|_{\mathrm{null}\,(T-\lambda_i I)} = \lambda_i I_{\mathrm{null}\,(T-\lambda_i I)}.$$

Corollary 11.3.2. *Let $T \in \mathcal{L}(V)$ be a normal operator, and denote by $\lambda_1, \ldots, \lambda_m$ the distinct eigenvalues of T.*

(1) $V = \mathrm{null}\,(T - \lambda_1 I) \oplus \cdots \oplus \mathrm{null}\,(T - \lambda_m I)$.
(2) If $i \neq j$, then $\mathrm{null}\,(T - \lambda_i I) \perp \mathrm{null}\,(T - \lambda_j I)$.

As we will see in the next section, we can use Corollary 11.3.2 to decompose the canonical matrix for a normal operator into a so-called "unitary diagonalization".

11.4 Applications of the Spectral Theorem: diagonalization

Let $e = (e_1, \ldots, e_n)$ be a basis for an n-dimensional vector space V, and let $T \in \mathcal{L}(V)$. In this section we denote the matrix $M(T)$ of T with respect to basis e by $[T]_e$. This is done to emphasize the dependency on the basis e. In other words, we have that

$$[Tv]_e = [T]_e [v]_e, \qquad \text{for all } v \in V,$$

where

$$[v]_e = \begin{bmatrix} v_1 \\ \vdots \\ v_n \end{bmatrix}$$

is the coordinate vector for $v = v_1 e_1 + \cdots + v_n e_n$ with $v_i \in \mathbb{F}$.

The operator T is diagonalizable if there exists a basis e such that $[T]_e$ is diagonal, i.e., if there exist $\lambda_1, \ldots, \lambda_n \in \mathbb{F}$ such that

$$[T]_e = \begin{bmatrix} \lambda_1 & & 0 \\ & \ddots & \\ 0 & & \lambda_n \end{bmatrix}.$$

The scalars $\lambda_1, \ldots, \lambda_n$ are necessarily eigenvalues of T, and $e_1, \ldots, e_n$ are the corresponding eigenvectors. We summarize this in the following proposition.

Proposition 11.4.1. *$T \in \mathcal{L}(V)$ is diagonalizable if and only if there exists a basis $(e_1, \ldots, e_n)$ consisting entirely of eigenvectors of T.*

We can reformulate this proposition using the change of basis transformations as follows. Suppose that e and f are bases of V such that $[T]_e$ is diagonal, and let S be the change of basis transformation such that $[v]_e = S[v]_f$. Then $S[T]_f S^{-1} = [T]_e$ is diagonal.

Proposition 11.4.2. *$T \in \mathcal{L}(V)$ is diagonalizable if and only if there exists an invertible matrix $S \in \mathbb{F}^{n \times n}$ such that*

$$S[T]_f S^{-1} = \begin{bmatrix} \lambda_1 & & 0 \\ & \ddots & \\ 0 & & \lambda_n \end{bmatrix},$$

where $[T]_f$ is the matrix for T with respect to a given arbitrary basis $f = (f_1, \ldots, f_n)$.

On the other hand, the Spectral Theorem tells us that T is diagonalizable with respect to an orthonormal basis if and only if T is normal. Recall that

$$[T^*]_f = [T]_f^*$$

for any orthonormal basis f of V. As before,

$$A^* = (\overline{a}_{ji})_{ij=1}^n, \qquad \text{for } A = (a_{ij})_{i,j=1}^n,$$

is the conjugate transpose of the matrix A. When $\mathbb{F} = \mathbb{R}$, note that $A^* = A^T$ is just the transpose of the matrix, where $A^T = (a_{ji})_{i,j=1}^n$.

The change of basis transformation between two orthonormal bases is called **unitary** in the complex case and **orthogonal** in the real case. Let $e = (e_1, \ldots, e_n)$ and $f = (f_1, \ldots, f_n)$ be two orthonormal bases of V, and let U be the change of basis matrix such that $[v]_f = U[v]_e$, for all $v \in V$. Then

$$\langle e_i, e_j \rangle = \delta_{ij} = \langle f_i, f_j \rangle = \langle Ue_i, Ue_j \rangle.$$

Since this holds for the basis e, it follows that U is unitary if and only if

$$\langle Uv, Uw \rangle = \langle v, w \rangle \quad \text{for all } v, w \in V. \tag{11.1}$$

This means that unitary matrices preserve the inner product. Operators that preserve the inner product are often also called **isometries**. Orthogonal matrices also define isometries.

By the definition of the adjoint, $\langle Uv, Uw \rangle = \langle v, U^*Uw \rangle$, and so Equation (11.1) implies that isometries are characterized by the property

$$U^*U = I, \qquad \text{for the unitary case,}$$
$$O^T O = I, \qquad \text{for the orthogonal case.}$$

The equation $U^*U = I$ implies that $U^{-1} = U^*$. For finite-dimensional inner product spaces, the left inverse of an operator is also the right inverse, and so

$$
\begin{aligned}
UU^* &= I \quad \text{if and only if} \quad U^*U = I, \\
OO^T &= I \quad \text{if and only if} \quad O^TO = I.
\end{aligned}
\tag{11.2}
$$

It is easy to see that the columns of a unitary matrix are the coefficients of the elements of an orthonormal basis with respect to another orthonormal basis. Therefore, the columns are orthonormal vectors in $\mathbb{C}^n$ (or in $\mathbb{R}^n$ in the real case). By Condition (11.2), this is also true for the rows of the matrix.

The Spectral Theorem tells us that $T \in \mathcal{L}(V)$ is normal if and only if $[T]_e$ is diagonal with respect to an orthonormal basis e for V, i.e., if there exists a unitary matrix U such that

$$
UTU^* = \begin{bmatrix} \lambda_1 & & 0 \\ & \ddots & \\ 0 & & \lambda_n \end{bmatrix}.
$$

Conversely, if a unitary matrix U exists such that $UTU^* = D$ is diagonal, then

$$
TT^* - T^*T = U^*(D\overline{D} - \overline{D}D)U = 0
$$

since diagonal matrices commute, and hence T is normal.

Let us summarize some of the definitions that we have seen in this section.

Definition 11.4.3. Given a square matrix $A \in \mathbb{F}^{n \times n}$, we call

(1) **symmetric** if $A = A^T$.
(2) **Hermitian** if $A = A^*$.
(3) **orthogonal** if $AA^T = I$.
(4) **unitary** if $AA^* = I$.

Note that every type of matrix in Definition 11.4.3 is an example of a normal operator. An example of a normal operator N that is neither Hermitian nor unitary is

$$
N = i \begin{bmatrix} -1 & -1 \\ -1 & 1 \end{bmatrix}.
$$

You can easily verify that $NN^* = N^*N$ and that iN is symmetric (not Hermitian).

Example 11.4.4. Consider the matrix

$$
A = \begin{bmatrix} 2 & 1+i \\ 1-i & 3 \end{bmatrix}
$$

from Example 11.1.5. To unitarily diagonalize A, we need to find a unitary matrix U and a diagonal matrix D such that $A = UDU^{-1}$. To do this, we need to first find a basis for $\mathbb{C}^2$ that consists entirely of orthonormal eigenvectors for the linear map $T \in \mathcal{L}(\mathbb{C}^2)$ defined by $Tv = Av$, for all $v \in \mathbb{C}^2$.

To find such an orthonormal basis, we start by finding the eigenspaces of T. We already determined that the eigenvalues of T are $\lambda_1 = 1$ and $\lambda_2 = 4$, so $D = \begin{bmatrix} 1 & 0 \\ 0 & 4 \end{bmatrix}$. It follows that

$$
\begin{aligned}
\mathbb{C}^2 &= \text{null}\,(T - I) \oplus \text{null}\,(T - 4I) \\
&= \text{span}((-1 - i, 1)) \oplus \text{span}((1 + i, 2)).
\end{aligned}
$$

Now apply the Gram-Schmidt procedure to each eigenspace in order to obtain the columns of U. Here,

$$
\begin{aligned}
A = UDU^{-1} &= \begin{bmatrix} \frac{-1-i}{\sqrt{3}} & \frac{1+i}{\sqrt{6}} \\ \frac{1}{\sqrt{3}} & \frac{2}{\sqrt{6}} \end{bmatrix} \begin{bmatrix} 1 & 0 \\ 0 & 4 \end{bmatrix} \begin{bmatrix} \frac{-1-i}{\sqrt{3}} & \frac{1+i}{\sqrt{6}} \\ \frac{1}{\sqrt{3}} & \frac{2}{\sqrt{6}} \end{bmatrix}^{-1} \\
&= \begin{bmatrix} \frac{-1-i}{\sqrt{3}} & \frac{1+i}{\sqrt{6}} \\ \frac{1}{\sqrt{3}} & \frac{2}{\sqrt{6}} \end{bmatrix} \begin{bmatrix} 1 & 0 \\ 0 & 4 \end{bmatrix} \begin{bmatrix} \frac{-1+i}{\sqrt{3}} & \frac{1}{\sqrt{3}} \\ \frac{1-i}{\sqrt{6}} & \frac{2}{\sqrt{6}} \end{bmatrix}.
\end{aligned}
$$

As an application, note that such diagonal decomposition allows us to easily compute powers and the exponential of matrices. Namely, if $A = UDU^{-1}$ with D diagonal, then we have

$$
A^n = (UDU^{-1})^n = UD^nU^{-1},
$$

$$
\exp(A) = \sum_{k=0}^{\infty} \frac{1}{k!} A^k = U \left(\sum_{k=0}^{\infty} \frac{1}{k!} D^k \right) U^{-1} = U \exp(D) U^{-1}.
$$

Example 11.4.5. Continuing Example 11.4.4,

$$
A^2 = (UDU^{-1})^2 = UD^2U^{-1} = U \begin{bmatrix} 1 & 0 \\ 0 & 16 \end{bmatrix} U^* = \begin{bmatrix} 6 & 5 + 5i \\ 5 - 5i & 11 \end{bmatrix},
$$

$$
A^n = (UDU^{-1})^n = UD^nU^{-1} = U \begin{bmatrix} 1 & 0 \\ 0 & 2^{2n} \end{bmatrix} U^* = \begin{bmatrix} \frac{2}{3}(1 + 2^{n-1}) & \frac{1+i}{3}(-1 + 2^{2n}) \\ \frac{1-i}{3}(-1 + 2^{2n}) & \frac{1}{3}(1 + 2^{2n+1}) \end{bmatrix},
$$

$$
\exp(A) = U \exp(D) U^{-1} = U \begin{bmatrix} e & 0 \\ 0 & e^4 \end{bmatrix} U^{-1} = \frac{1}{3} \begin{bmatrix} 2e + e^4 & e^4 - e + i(e^4 - e) \\ e^4 - e + i(e - e^4) & e + 2e^4 \end{bmatrix}.
$$

11.5   Positive operators

Recall that self-adjoint operators are the operator analog for real numbers. Let us now define the operator analog for positive (or, more precisely, non-negative) real numbers.

Definition 11.5.1. An operator $T \in \mathcal{L}(V)$ is called **positive** (denoted $T \geq 0$) if $T = T^*$ and $\langle Tv, v \rangle \geq 0$ for all $v \in V$.

(If V is a complex vector space, then the condition of self-adjointness follows from the condition $\langle Tv, v \rangle \geq 0$ and hence can be dropped.)

Example 11.5.2. Note that, for all $T \in \mathcal{L}(V)$, we have $T^*T \geq 0$ since T^*T is self-adjoint and $\langle T^*Tv, v \rangle = \langle Tv, Tv \rangle \geq 0$.

Example 11.5.3. Let $U \subset V$ be a subspace of V and P_U be the orthogonal projection onto U. Then $P_U \geq 0$. To see this, write $V = U \oplus U^\perp$ and $v = u_v + u_v^\perp$ for each $v \in V$, where $u_v \in U$ and $u_v^\perp \in U^\perp$. Then $\langle P_U v, w \rangle = \langle u_v, u_w + u_w^\perp \rangle = \langle u_v, u_w \rangle = \langle u_v + u_v^\perp, u_w \rangle = \langle v, P_U w \rangle$ so that $P_U^* = P_U$. Also, setting $v = w$ in the above string of equations, we obtain $\langle P_U v, v \rangle = \langle u_v, u_v \rangle \geq 0$, for all $v \in V$. Hence, $P_U \geq 0$.

If λ is an eigenvalue of a positive operator T and $v \in V$ is an associated eigenvector, then $\langle Tv, v \rangle = \langle \lambda v, v \rangle = \lambda \langle v, v \rangle \geq 0$. Since $\langle v, v \rangle \geq 0$ for all vectors $v \in V$, it follows that $\lambda \geq 0$. This fact can be used to define $\sqrt{T}$ by setting

$$\sqrt{T} e_i = \sqrt{\lambda_i} e_i,$$

where λ_i are the eigenvalues of T with respect to the orthonormal basis $e = (e_1, \ldots, e_n)$. We know that these exist by the Spectral Theorem.

11.6 Polar decomposition

Continuing the analogy between $\mathbb{C}$ and $\mathcal{L}(V)$, recall the polar form of a complex number $z = |z|e^{i\theta}$, where $|z|$ is the absolute value or modulus of z and $e^{i\theta}$ lies on the unit circle in $\mathbb{R}^2$. In terms of an operator $T \in \mathcal{L}(V)$, where V is a complex inner product space, a unitary operator U takes the role of $e^{i\theta}$, and $|T|$ takes the role of the modulus. As in Section 11.5, $T^*T \geq 0$ so that $|T| := \sqrt{T^*T}$ exists and satisfies $|T| \geq 0$ as well.

Theorem 11.6.1. *For each $T \in \mathcal{L}(V)$, there exists a unitary U such that*

$$T = U|T|.$$

*This is called the **polar decomposition** of T.*

Sketch of proof. We start by noting that

$$\|Tv\|^2 = \| \, |T| \, v\|^2,$$

since $\langle Tv, Tv \rangle = \langle v, T^*Tv \rangle = \langle \sqrt{T^*T}v, \sqrt{T^*T}v \rangle$. This implies that $\mathrm{null}(T) = \mathrm{null}(|T|)$. By the Dimension Formula, this also means that $\dim(\mathrm{range}(T)) = \dim(\mathrm{range}(|T|))$. Moreover, we can define an isometry $S : \mathrm{range}(|T|) \to \mathrm{range}(T)$ by setting

$$S(|T|v) = Tv.$$

The trick is now to define a unitary operator U on all of V such that the restriction of U onto the range of $|T|$ is S, i.e.,

$$U|_{\mathrm{range}(|T|)} = S.$$

Note that $\mathrm{null}\,(|T|)\perp\mathrm{range}\,(|T|)$, i.e., for $v\in\mathrm{null}\,(|T|)$ and $w=|T|u\in\mathrm{range}\,(|T|)$,

$$\langle w,v\rangle=\langle|T|u,v\rangle=\langle u,|T|v\rangle=\langle u,0\rangle=0$$

since $|T|$ is self-adjoint.

Pick an orthonormal basis $e=(e_1,\ldots,e_m)$ of $\mathrm{null}\,(|T|)$ and an orthonormal basis $f=(f_1,\ldots,f_m)$ of $(\mathrm{range}\,(T))^\perp$. Set $\tilde{S}e_i=f_i$, and extend $\tilde{S}$ to all of $\mathrm{null}\,(|T|)$ by linearity. Since $\mathrm{null}\,(|T|)\perp\mathrm{range}\,(|T|)$, any $v\in V$ can be uniquely written as $v=v_1+v_2$, where $v_1\in\mathrm{null}\,(|T|)$ and $v_2\in\mathrm{range}\,(|T|)$. Now define $U:V\to V$ by setting $Uv=\tilde{S}v_1+Sv_2$. Then U is an isometry. Moreover, U is also unitary, as shown by the following calculation application of the Pythagorean theorem:

$$\|Uv\|^2=\|\tilde{S}v_1+Sv_2\|^2=\|\tilde{S}v_1\|^2+\|Sv_2\|^2$$
$$=\|v_1\|^2+\|v_2\|^2=\|v\|^2.$$

Also, note that $U|T|=T$ by construction since $U|_{\mathrm{null}\,(|T|)}$ is irrelevant. $\qquad\square$

11.7 Singular-value decomposition

The singular-value decomposition generalizes the notion of diagonalization. To unitarily diagonalize $T\in\mathcal{L}(V)$ means to find an orthonormal basis e such that T is diagonal with respect to this basis, i.e.,

$$M(T;e,e)=[T]_e=\begin{bmatrix}\lambda_1 & & 0\\ & \ddots & \\ 0 & & \lambda_n\end{bmatrix},$$

where the notation $M(T;e,e)$ indicates that the basis e is used both for the domain and codomain of T. The Spectral Theorem tells us that unitary diagonalization can only be done for normal operators. In general, we can find two orthonormal bases e and f such that

$$M(T;e,f)=\begin{bmatrix}s_1 & & 0\\ & \ddots & \\ 0 & & s_n\end{bmatrix},$$

which means that $Te_i=s_if_i$ even if T is not normal. The scalars s_i are called **singular values** of T. If T is diagonalizable, then these are the absolute values of the eigenvalues.

Theorem 11.7.1. *All $T\in\mathcal{L}(V)$ have a singular-value decomposition. That is, there exist orthonormal bases $e=(e_1,\ldots,e_n)$ and $f=(f_1,\ldots,f_n)$ such that*

$$Tv=s_1\langle v,e_1\rangle f_1+\cdots+s_n\langle v,e_n\rangle f_n,$$

where s_i are the singular values of T.

Proof. Since $|T| \geq 0$, it is also self-adjoint. Thus, by the Spectral Theorem, there is an orthonormal basis $e = (e_1, \ldots, e_n)$ for V such that $|T|e_i = s_i e_i$. Let U be the unitary matrix in the polar decomposition of T. Since e is orthonormal, we can write any vector $v \in V$ as

$$v = \langle v, e_1 \rangle e_1 + \cdots + \langle v, e_n \rangle e_n,$$

and hence

$$Tv = U|T|v = s_1 \langle v, e_1 \rangle U e_1 + \cdots + s_n \langle v, e_n \rangle U e_n.$$

Now set $f_i = U e_i$ for all $1 \leq i \leq n$. Since U is unitary, $(f_1, \ldots, f_n)$ is also an orthonormal basis, proving the theorem. $\square$

Exercises for Chapter 11

Calculational Exercises

(1) Consider $\mathbb{R}^3$ with two orthonormal bases: the canonical basis $e = (e_1, e_2, e_3)$ and the basis $f = (f_1, f_2, f_3)$, where

$$f_1 = \frac{1}{\sqrt{3}}(1,1,1), \ f_2 = \frac{1}{\sqrt{6}}(1,-2,1), \ f_3 = \frac{1}{\sqrt{2}}(1,0,-1).$$

Find the canonical matrix, A, of the linear map $T \in \mathcal{L}(\mathbb{R}^3)$ with eigenvectors f_1, f_2, f_3 and eigenvalues $1, 1/2, -1/2$, respectively.

(2) For each of the following matrices, verify that A is Hermitian by showing that $A = A^*$, find a unitary matrix U such that $U^{-1}AU$ is a diagonal matrix, and compute $\exp(A)$.

(a) $A = \begin{bmatrix} 4 & 1-i \\ 1+i & 5 \end{bmatrix}$ (b) $A = \begin{bmatrix} 3 & -i \\ i & 3 \end{bmatrix}$ (c) $A = \begin{bmatrix} 6 & 2+2i \\ 2-2i & 4 \end{bmatrix}$

(d) $A = \begin{bmatrix} 0 & 3+i \\ 3-i & -3 \end{bmatrix}$ (e) $A = \begin{bmatrix} 5 & 0 & 0 \\ 0 & -1 & -1+i \\ 0 & -1-i & 0 \end{bmatrix}$ (f) $A = \begin{bmatrix} 2 & \frac{i}{\sqrt{2}} & \frac{-i}{\sqrt{2}} \\ \frac{-i}{\sqrt{2}} & 2 & 0 \\ \frac{i}{\sqrt{2}} & 0 & 2 \end{bmatrix}$

(3) For each of the following matrices, either find a matrix P (not necessarily unitary) such that $P^{-1}AP$ is a diagonal matrix, or show why no such matrix exists.

(a) $A = \begin{bmatrix} 19 & -9 & -6 \\ 25 & -11 & -9 \\ 17 & -9 & -4 \end{bmatrix}$ (b) $A = \begin{bmatrix} -1 & 4 & -2 \\ -3 & 4 & 0 \\ -3 & 1 & 3 \end{bmatrix}$ (c) $A = \begin{bmatrix} 5 & 0 & 0 \\ 1 & 5 & 0 \\ 0 & 1 & 5 \end{bmatrix}$

(d) $A = \begin{bmatrix} 0 & 0 & 0 \\ 0 & 0 & 0 \\ 3 & 0 & 1 \end{bmatrix}$ (e) $A = \begin{bmatrix} -i & 1 & 1 \\ -i & 1 & 1 \\ -i & 1 & 1 \end{bmatrix}$ (f) $A = \begin{bmatrix} 0 & 0 & i \\ 4 & 0 & i \\ 0 & 0 & i \end{bmatrix}$

(4) Let $r \in \mathbb{R}$ and let $T \in \mathcal{L}(\mathbb{C}^2)$ be the linear map with canonical matrix

$$T = \begin{pmatrix} 1 & -1 \\ -1 & r \end{pmatrix}.$$

   (a) Find the eigenvalues of T.
   (b) Find an orthonormal basis of $\mathbb{C}^2$ consisting of eigenvectors of T.
   (c) Find a unitary matrix U such that UTU^* is diagonal.

(5) Let A be the complex matrix given by:

$$A = \begin{bmatrix} 5 & 0 & 0 \\ 0 & -1 & -1+i \\ 0 & -1-i & 0 \end{bmatrix}$$

   (a) Find the eigenvalues of A.
   (b) Find an orthonormal basis of eigenvectors of A.
   (c) Calculate $|A| = \sqrt{A^*A}$.
   (d) Calculate e^A.

(6) Let $\theta \in \mathbb{R}$, and let $T \in \mathcal{L}(\mathbb{C}^2)$ have canonical matrix

$$M(T) = \begin{pmatrix} 1 & e^{i\theta} \\ e^{-i\theta} & -1 \end{pmatrix}.$$

   (a) Find the eigenvalues of T.
   (b) Find an orthonormal basis for $\mathbb{C}^2$ that consists of eigenvectors for T.

Proof-Writing Exercises

(1) Prove or give a counterexample: The product of any two self-adjoint operators on a finite-dimensional vector space is self-adjoint.
(2) Prove or give a counterexample: Every unitary matrix is invertible.
(3) Let V be a finite-dimensional vector space over $\mathbb{F}$, and suppose that $T \in \mathcal{L}(V)$ satisfies $T^2 = T$. Prove that T is an orthogonal projection if and only if T is self-adjoint.
(4) Let V be a finite-dimensional inner product space over $\mathbb{C}$, and suppose that $T \in \mathcal{L}(V)$ has the property that $T^* = -T$. (We call T a **skew Hermitian** operator on V.)

   (a) Prove that the operator $iT \in \mathcal{L}(V)$ defined by $(iT)(v) = i(T(v))$, for each $v \in V$, is Hermitian.
   (b) Prove that the canonical matrix for T can be unitarily diagonalized.
   (c) Prove that T has purely imaginary eigenvalues.

(5) Let V be a finite-dimensional vector space over $\mathbb{F}$, and suppose that $S, T \in \mathcal{L}(V)$ are positive operators on V. Prove that $S+T$ is also a positive operator on T.
(6) Let V be a finite-dimensional vector space over $\mathbb{F}$, and let $T \in \mathcal{L}(V)$ be any operator on V. Prove that T is invertible if and only if 0 is not a singular value of T.

Appendix A

Supplementary Notes on Matrices and Linear Systems

As discussed in Chapter 1, there are many ways in which you might try to solve a system of linear equation involving a finite number of variables. These supplementary notes are intended to illustrate the use of Linear Algebra in solving such systems. In particular, any arbitrary number of equations in any number of unknowns — as long as both are finite — can be encoded as a single **matrix equation**. As you will see, this has many computational advantages, but, perhaps more importantly, it also allows us to better understand linear systems abstractly. Specifically, by exploiting the deep connection between matrices and so-called linear maps, one can completely determine all possible solutions to any linear system.

These notes are also intended to provide a self-contained introduction to matrices and important matrix operations. As you read the sections below, remember that a matrix is, in general, nothing more than a rectangular array of real or complex numbers. Matrices *are not* linear maps. Instead, a matrix can (and will often) be used to *define* a linear map.

A.1 From linear systems to matrix equations

We begin this section by reviewing the definition of and notation for matrices. We then review several different conventions for denoting and studying systems of linear equations. This point of view has a long history of exploration, and numerous computational devices — including several computer programming languages — have been developed and optimized specifically for analyzing matrix equations.

A.1.1 *Definition of and notation for matrices*

Let $m, n \in \mathbb{Z}_+$ be positive integers, and, as usual, let $\mathbb{F}$ denote either $\mathbb{R}$ or $\mathbb{C}$. Then we begin by defining an $m \times n$ **matrix** A to be a rectangular array of numbers

$$A = (a_{ij})_{i,j=1}^{m,n} = (A^{(i,j)})_{i,j=1}^{m,n} = \left.\begin{bmatrix} a_{11} & \cdots & a_{1n} \\ \vdots & \ddots & \vdots \\ a_{m1} & \cdots & a_{mn} \end{bmatrix}\right\} m \text{ numbers}$$

$$\underbrace{}_{n \text{ numbers}}$$

where each element $a_{ij} \in \mathbb{F}$ in the array is called an **entry** of A (specifically, a_{ij} is called the "i, j entry"). We say that i indexes the **rows** of A as it ranges over the set $\{1, \ldots, m\}$ and that j indexes the **columns** of A as it ranges over the set $\{1, \ldots, n\}$. We also say that the matrix A has **size** $m \times n$ and note that it is a (finite) sequence of doubly-subscripted numbers for which the two subscripts in no way depend upon each other.

Definition A.1.1. Given positive integers $m, n \in \mathbb{Z}_+$, we use $\mathbb{F}^{m \times n}$ to denote the set of all $m \times n$ matrices having entries in $\mathbb{F}$.

Example A.1.2. The matrix $A = \begin{bmatrix} 1 & 0 & 2 \\ -1 & 3 & i \end{bmatrix} \in \mathbb{C}^{2 \times 3}$, but $A \notin \mathbb{R}^{2 \times 3}$ since the "2, 3" entry of A is not in $\mathbb{R}$.

Given the ubiquity of matrices in both abstract and applied mathematics, a rich vocabulary has been developed for describing various properties and features of matrices. In addition, there is also a rich set of equivalent notations. For the purposes of these notes, we will use the above notation unless the size of the matrix is understood from the context or is unimportant. In this case, we will drop much of this notation and denote a matrix simply as

$$A = (a_{ij}) \quad \text{or} \quad A = (a_{ij})_{m \times n}.$$

To get a sense of the essential vocabulary, suppose that we have an $m \times n$ matrix $A = (a_{ij})$ with $m = n$. Then we call A a **square** matrix. The elements $a_{11}, a_{22}, \ldots, a_{nn}$ in a square matrix form the **main diagonal** of A, and the elements $a_{1n}, a_{2,n-1}, \ldots, a_{n1}$ form what is sometimes called the **skew main diagonal** of A. Entries not on the main diagonal are also often called **off-diagonal** entries, and a matrix whose off-diagonal entries are all zero is called a **diagonal matrix**. It is common to call $a_{12}, a_{23}, \ldots, a_{n-1,n}$ the **superdiagonal** of A and $a_{21}, a_{32}, \ldots, a_{n,n-1}$ the **subdiagonal** of A. The motivation for this terminology should be clear if you create a sample square matrix and trace the entries within these particular subsequences of the matrix.

Square matrices are important because they are fundamental to applications of Linear Algebra. In particular, virtually every use of Linear Algebra either involves square matrices directly or employs them in some indirect manner. In addition, virtually every usage also involves the notion of **vector**, by which we mean here either an $m \times 1$ matrix (a.k.a. a **row vector**) or a $1 \times n$ matrix (a.k.a. a **column vector**).

Example A.1.3. Suppose that $A = (a_{ij})$, $B = (b_{ij})$, $C = (c_{ij})$, $D = (d_{ij})$, and $E = (e_{ij})$ are the following matrices over $\mathbb{F}$:

$$A = \begin{bmatrix} 3 \\ -1 \\ 1 \end{bmatrix}, \ B = \begin{bmatrix} 4 & -1 \\ 0 & 2 \end{bmatrix}, \ C = \begin{bmatrix} 1, 4, 2 \end{bmatrix}, \ D = \begin{bmatrix} 1 & 5 & 2 \\ -1 & 0 & 1 \\ 3 & 2 & 4 \end{bmatrix}, \ E = \begin{bmatrix} 6 & 1 & 3 \\ -1 & 1 & 2 \\ 4 & 1 & 3 \end{bmatrix}.$$

Then we say that A is a 3×1 matrix (a.k.a. a column vector), B is a 2×2 square matrix, C is a 1×3 matrix (a.k.a. a row vector), and both D and E are square 3×3 matrices. Moreover, only B is an upper-triangular matrix (as defined below), and none of the matrices in this example are diagonal matrices.

We can discuss individual entries in each matrix. E.g.,

- the 2nd row of D is $d_{21} = -1$, $d_{22} = 0$, and $d_{23} = 1$.
- the main diagonal of D is the sequence $d_{11} = 1, d_{22} = 0, d_{33} = 4$.
- the skew main diagonal of D is the sequence $d_{13} = 2, d_{22} = 0, d_{31} = 3$.
- the off-diagonal entries of D are (by row) $d_{12}, d_{13}, d_{21}, d_{23}, d_{31}$, and d_{32}.
- the 2nd column of E is $e_{12} = e_{22} = e_{32} = 1$.
- the superdiagonal of E is the sequence $e_{12} = 1, e_{23} = 2$.
- the subdiagonal of E is the sequence $e_{21} = -1, e_{32} = 1$.

A square matrix $A = (a_{ij}) \in \mathbb{F}^{n \times n}$ is called **upper triangular** (resp. **lower triangular**) if $a_{ij} = 0$ for each pair of integers $i, j \in \{1, \ldots, n\}$ such that $i > j$ (resp. $i < j$). In other words, A is triangular if it has the form

$$\begin{bmatrix} a_{11} & a_{12} & a_{13} & \cdots & a_{1n} \\ 0 & a_{22} & a_{23} & \cdots & a_{2n} \\ 0 & 0 & a_{33} & \cdots & a_{3n} \\ \vdots & \vdots & \vdots & \ddots & \vdots \\ 0 & 0 & 0 & \cdots & a_{nn} \end{bmatrix} \quad \text{or} \quad \begin{bmatrix} a_{11} & 0 & 0 & \cdots & 0 \\ a_{21} & a_{22} & 0 & \cdots & 0 \\ a_{31} & a_{32} & a_{33} & \cdots & 0 \\ \vdots & \vdots & \vdots & \ddots & \vdots \\ a_{n1} & a_{n2} & a_{n3} & \cdots & a_{nn} \end{bmatrix}.$$

Note that a diagonal matrix is simultaneously both an upper triangular matrix and a lower triangular matrix.

Two particularly important examples of diagonal matrices are defined as follows: Given any positive integer $n \in \mathbb{Z}_+$, we can construct the **identity matrix** I_n and the **zero matrix** $0_{n \times n}$ by setting

$$I_n = \begin{bmatrix} 1 & 0 & 0 & \cdots & 0 & 0 \\ 0 & 1 & 0 & \cdots & 0 & 0 \\ 0 & 0 & 1 & \cdots & 0 & 0 \\ \vdots & \vdots & \vdots & \ddots & \vdots & \vdots \\ 0 & 0 & 0 & \cdots & 1 & 0 \\ 0 & 0 & 0 & \cdots & 0 & 1 \end{bmatrix} \quad \text{and} \quad 0_{n \times n} = \begin{bmatrix} 0 & 0 & 0 & \cdots & 0 & 0 \\ 0 & 0 & 0 & \cdots & 0 & 0 \\ 0 & 0 & 0 & \cdots & 0 & 0 \\ \vdots & \vdots & \vdots & \ddots & \vdots & \vdots \\ 0 & 0 & 0 & \cdots & 0 & 0 \\ 0 & 0 & 0 & \cdots & 0 & 0 \end{bmatrix},$$

where each of these matrices is understood to be a square matrix of size $n \times n$. The
zero matrix $0_{m \times n}$ is analogously defined for any $m, n \in \mathbb{Z}_+$ and has size $m \times n$. I.e.,

$$
0_{m \times n} = \left.\begin{bmatrix} 0\,0\,0 \cdots 0\,0 \\ 0\,0\,0 \cdots 0\,0 \\ 0\,0\,0 \cdots 0\,0 \\ \vdots\,\vdots\,\vdots\,\ddots\,\vdots\,\vdots \\ 0\,0\,0 \cdots 0\,0 \\ 0\,0\,0 \cdots 0\,0 \end{bmatrix}\right\} m \;\; \text{rows}
$$

$$
\underbrace{}_{n \;\; \text{columns}}
$$

A.1.2 *Using matrices to encode linear systems*

Let $m, n \in \mathbb{Z}_+$ be positive integers. Then a **system of m linear equations in n
unknowns** $x_1, \ldots, x_n$ looks like

$$
\left.\begin{aligned}
a_{11}x_1 + a_{12}x_2 + a_{13}x_3 + \cdots + a_{1n}x_n &= b_1 \\
a_{21}x_1 + a_{22}x_2 + a_{23}x_3 + \cdots + a_{2n}x_n &= b_2 \\
a_{31}x_1 + a_{32}x_2 + a_{33}x_3 + \cdots + a_{3n}x_n &= b_3 \\
&\;\;\vdots \\
a_{m1}x_1 + a_{m2}x_2 + a_{m3}x_3 + \cdots + a_{mn}x_n &= b_m
\end{aligned}\right\} , \tag{A.1}
$$

where each $a_{ij}, b_i \in \mathbb{F}$ is a scalar for $i = 1, 2, \ldots, m$ and $j = 1, 2, \ldots, n$. In other
words, each scalar $b_1, \ldots, b_m \in \mathbb{F}$ is being written as a linear combination of the
unknowns $x_1, \ldots, x_n$ using coefficients from the field $\mathbb{F}$. To **solve** System (A.1)
means to describe the set of all possible values for $x_1, \ldots, x_n$ (when thought of
as scalars in $\mathbb{F}$) such that each of the m equations in System (A.1) is satisfied
simultaneously.

Rather than dealing directly with a given linear system, it is often convenient to
first encode the system using less cumbersome notation. Specifically, System (A.1)
can be summarized using exactly three matrices. First, we collect the coefficients
from each equation into the $m \times n$ matrix $A = (a_{ij}) \in \mathbb{F}^{m \times n}$, which we call the
coefficient matrix for the linear system. Similarly, we assemble the unknowns
$x_1, x_2, \ldots, x_n$ into an $n \times 1$ column vector $x = (x_i) \in \mathbb{F}^n$, and the right-hand sides
$b_1, b_2, \ldots, b_m$ of the equation are used to form an $m \times 1$ column vector $b = (b_i) \in \mathbb{F}^m$.
In other words,

$$
A = \begin{bmatrix} a_{11} & a_{12} & \cdots & a_{1n} \\ a_{21} & a_{22} & \cdots & a_{2n} \\ \vdots & \vdots & \ddots & \vdots \\ a_{m1} & a_{m2} & \cdots & a_{mn} \end{bmatrix}, \quad x = \begin{bmatrix} x_1 \\ x_2 \\ \vdots \\ x_n \end{bmatrix}, \quad \text{and} \quad b = \begin{bmatrix} b_1 \\ b_2 \\ \vdots \\ b_m \end{bmatrix}.
$$

Then the left-hand side of the ith equation in System (A.1) can be recovered by taking the **dot product** (a.k.a. **Euclidean inner product**) of x with the ith row in A:

$$\begin{bmatrix} a_{i1} \; a_{i2} \; \cdots \; a_{in} \end{bmatrix} \cdot x = \sum_{j=1}^{n} a_{ij} x_j = a_{i1}x_1 + a_{i2}x_2 + a_{i3}x_3 + \cdots + a_{in}x_n.$$

In general, we can extend the dot product between two vectors in order to form the product of any two matrices (as in Section A.2.2). For the purposes of this section, though, it suffices to define the product of the matrix $A \in \mathbb{F}^{m \times n}$ and the vector $x \in \mathbb{F}^n$ to be

$$Ax = \begin{bmatrix} a_{11} & a_{12} & \cdots & a_{1n} \\ a_{21} & a_{22} & \cdots & a_{2n} \\ \vdots & \vdots & \ddots & \vdots \\ a_{m1} & a_{m2} & \cdots & a_{mn} \end{bmatrix} \begin{bmatrix} x_1 \\ x_2 \\ \vdots \\ x_n \end{bmatrix} = \begin{bmatrix} a_{11}x_1 + a_{12}x_2 + \cdots + a_{1n}x_n \\ a_{21}x_1 + a_{22}x_2 + \cdots + a_{2n}x_n \\ \vdots \\ a_{m1}x_1 + a_{m2}x_2 + \cdots + a_{mn}x_n \end{bmatrix}. \tag{A.2}$$

Then, since each entry in the resulting $m \times 1$ column vector $Ax \in \mathbb{F}^m$ corresponds exactly to the left-hand side of each equation in System (A.1), we have effectively encoded System (A.1) as the single **matrix equation**

$$Ax = \begin{bmatrix} a_{11}x_1 + a_{12}x_2 + \cdots + a_{1n}x_n \\ a_{21}x_1 + a_{22}x_2 + \cdots + a_{2n}x_n \\ \vdots \\ a_{m1}x_1 + a_{m2}x_2 + \cdots + a_{mn}x_n \end{bmatrix} = \begin{bmatrix} b_1 \\ \vdots \\ b_m \end{bmatrix} = b. \tag{A.3}$$

Example A.1.4. The linear system

$$\left. \begin{aligned} x_1 + 6x_2 \qquad\quad + 4x_5 - 2x_6 &= 14 \\ x_3 \quad + 3x_5 + \; x_6 &= -3 \\ x_4 + 5x_5 + 2x_6 &= 11 \end{aligned} \right\}$$

has three equations and involves the six variables $x_1, x_2, \ldots, x_6$. One can check that possible solutions to this system include

$$\begin{bmatrix} x_1 \\ x_2 \\ x_3 \\ x_4 \\ x_6 \\ x_6 \end{bmatrix} = \begin{bmatrix} 14 \\ 0 \\ -3 \\ 11 \\ 0 \\ 0 \end{bmatrix} \quad \text{and} \quad \begin{bmatrix} x_1 \\ x_2 \\ x_3 \\ x_4 \\ x_6 \\ x_6 \end{bmatrix} = \begin{bmatrix} 6 \\ 1 \\ -9 \\ -5 \\ 2 \\ 3 \end{bmatrix}.$$

Note that, in describing these solutions, we have used the six unknowns $x_1, x_2, \ldots, x_6$ to form the 6×1 column vector $x = (x_i) \in \mathbb{F}^6$. We can similarly form the coefficient matrix $A \in \mathbb{F}^{3 \times 6}$ and the 3×1 column vector $b \in \mathbb{F}^3$, where

$$A = \begin{bmatrix} 1 & 6 & 0 & 0 & 4 & -2 \\ 0 & 0 & 1 & 0 & 3 & 1 \\ 0 & 0 & 0 & 1 & 5 & 2 \end{bmatrix} \quad \text{and} \quad \begin{bmatrix} b_1 \\ b_2 \\ b_3 \end{bmatrix} = \begin{bmatrix} 14 \\ -3 \\ 11 \end{bmatrix}.$$

You should check that, given these matrices, each of the solutions given above satisfies Equation (A.3).

We close this section by mentioning another common convention for encoding linear systems. Specifically, rather than attempting to solve Equation (A.1) directly, one can instead look at the equivalent problem of describing all coefficients $x_1, \ldots, x_n \in \mathbb{F}$ for which the following **vector equation** is satisfied:

$$x_1 \begin{bmatrix} a_{11} \\ a_{21} \\ a_{31} \\ \vdots \\ a_{m1} \end{bmatrix} + x_2 \begin{bmatrix} a_{12} \\ a_{22} \\ a_{32} \\ \vdots \\ a_{m2} \end{bmatrix} + x_3 \begin{bmatrix} a_{13} \\ a_{23} \\ a_{33} \\ \vdots \\ a_{m3} \end{bmatrix} + \cdots + x_n \begin{bmatrix} a_{1n} \\ a_{2n} \\ a_{3n} \\ \vdots \\ a_{mn} \end{bmatrix} = \begin{bmatrix} b_1 \\ b_2 \\ b_3 \\ \vdots \\ b_m \end{bmatrix}. \tag{A.4}$$

This approach emphasizes analysis of the so-called **column vectors** $A^{(\cdot,j)}$ ($j = 1, \ldots, n$) of the coefficient matrix A in the matrix equation $Ax = b$. (See in Section A.2 for more details about how Equation (A.4) is formed.) Conversely, it is also common to directly encounter Equation (A.4) when studying certain questions about vectors in $\mathbb{F}^n$.

It is important to note that System (A.1) differs from Equations (A.3) and (A.4) only in terms of notation. The common aspect of these different representations is that the left-hand side of each equation in System (A.1) is a linear sum. Because of this, it is also common to rewrite System (A.1) using more compact notation such as

$$\sum_{k=1}^n a_{1k}x_k = b_1, \ \sum_{k=1}^n a_{2k}x_k = b_2, \ \sum_{k=1}^n a_{3k}x_k = b_3, \ \ldots, \ \sum_{k=1}^n a_{mk}x_k = b_m.$$

A.2 Matrix arithmetic

In this section, we examine algebraic properties of the set $\mathbb{F}^{m \times n}$ (where $m, n \in \mathbb{Z}_+$). Specifically, $\mathbb{F}^{m \times n}$ forms a vector space under the operations of component-wise addition and scalar multiplication, and it is isomorphic to $\mathbb{F}^{mn}$ as a vector space.

We also define a multiplication operation between matrices of compatible size and show that this multiplication operation interacts with the vector space structure on $\mathbb{F}^{m \times n}$ in a natural way. In particular, $\mathbb{F}^{n \times n}$ forms an algebra over $\mathbb{F}$ with respect to these operations. (See Section C.3 for the definition of an algebra.)

A.2.1 *Addition and scalar multiplication*

Let $A = (a_{ij})$ and $B = (b_{ij})$ be $m \times n$ matrices over $\mathbb{F}$ (where $m, n \in \mathbb{Z}_+$), and let $\alpha \in \mathbb{F}$. Then **matrix addition** $A + B = ((a+b)_{ij})_{m \times n}$ and **scalar multiplication** $\alpha A = ((\alpha a)_{ij})_{m \times n}$ are both defined component-wise, meaning

$$(a+b)_{ij} = a_{ij} + b_{ij} \ \text{ and } \ (\alpha a)_{ij} = \alpha a_{ij}.$$

Equivalently, $A + B$ is the $m \times n$ matrix given by

$$\begin{bmatrix} a_{11} & \cdots & a_{1n} \\ \vdots & \ddots & \vdots \\ a_{m1} & \cdots & a_{mn} \end{bmatrix} + \begin{bmatrix} b_{11} & \cdots & b_{1n} \\ \vdots & \ddots & \vdots \\ b_{m1} & \cdots & b_{mn} \end{bmatrix} = \begin{bmatrix} a_{11} + b_{11} & \cdots & a_{1n} + b_{1n} \\ \vdots & \ddots & \vdots \\ a_{m1} + b_{m1} & \cdots & a_{mn} + b_{mn} \end{bmatrix},$$

and αA is the $m \times n$ matrix given by

$$\alpha \begin{bmatrix} a_{11} & \cdots & a_{1n} \\ \vdots & \ddots & \vdots \\ a_{m1} & \cdots & a_{mn} \end{bmatrix} = \begin{bmatrix} \alpha a_{11} & \cdots & \alpha a_{1n} \\ \vdots & \ddots & \vdots \\ \alpha a_{m1} & \cdots & \alpha a_{mn} \end{bmatrix}.$$

Example A.2.1. With notation as in Example A.1.3,

$$D + E = \begin{bmatrix} 7 & 6 & 5 \\ -2 & 1 & 3 \\ 7 & 3 & 7 \end{bmatrix},$$

and no two other matrices from Example A.1.3 can be added since their sizes are not compatible. Similarly, we can make calculations like

$$D - E = D + (-1)E = \begin{bmatrix} -5 & 4 & -1 \\ 0 & -1 & -1 \\ -1 & 1 & 1 \end{bmatrix} \quad \text{and} \quad 0D = 0E = \begin{bmatrix} 0 & 0 & 0 \\ 0 & 0 & 0 \\ 0 & 0 & 0 \end{bmatrix} = 0_{3 \times 3}.$$

It is important to note that the above operations endow $\mathbb{F}^{m \times n}$ with a natural vector space structure. As a vector space, $\mathbb{F}^{m \times n}$ is seen to have dimension mn since we can build the **standard basis matrices**

$$E_{11}, E_{12}, \ldots, E_{1n}, E_{21}, E_{22}, \ldots, E_{2n}, \ldots, E_{m1}, E_{m2}, \ldots, E_{mn}$$

by analogy to the standard basis for $\mathbb{F}^m$. That is, each $E_{k\ell} = ((e^{(k,\ell)})_{ij})$ satisfies

$$(e^{(k,\ell)})_{ij} = \begin{cases} 1, & \text{if } i = k \text{ and } j = \ell \\ 0, & \text{otherwise} \end{cases}.$$

This allows us to build a vector space isomorphism $\mathbb{F}^{m \times n} \to \mathbb{F}^{mn}$ using a bijection that simply "lays each matrix out flat". In other words, given $A = (a_{ij}) \in \mathbb{F}^{m \times n}$,

$$\begin{bmatrix} a_{11} & \cdots & a_{1n} \\ \vdots & \ddots & \vdots \\ a_{m1} & \cdots & a_{mn} \end{bmatrix} \mapsto (a_{11}, a_{12}, \ldots, a_{1n}, a_{21}, a_{22}, \ldots, a_{2n}, \ldots, a_{m1}, a_{m2}, \ldots, a_{mn}) \in \mathbb{F}^{mn}.$$

Example A.2.2. The vector space $\mathbb{R}^{2 \times 3}$ of 2×3 matrices over $\mathbb{R}$ has standard basis

$$\left\{ E_{11} = \begin{bmatrix} 1 & 0 & 0 \\ 0 & 0 & 0 \end{bmatrix}, E_{12} = \begin{bmatrix} 0 & 1 & 0 \\ 0 & 0 & 0 \end{bmatrix}, E_{13} = \begin{bmatrix} 0 & 0 & 1 \\ 0 & 0 & 0 \end{bmatrix}, \right.$$

$$\left. E_{21} = \begin{bmatrix} 0 & 0 & 0 \\ 1 & 0 & 0 \end{bmatrix}, E_{22} = \begin{bmatrix} 0 & 0 & 0 \\ 0 & 1 & 0 \end{bmatrix}, E_{23} = \begin{bmatrix} 0 & 0 & 0 \\ 0 & 0 & 1 \end{bmatrix} \right\},$$

which is seen to naturally correspond with the standard basis $\{e_1, \ldots, e_6\}$ for $\mathbb{R}_6$, where

$$e_1 = (1, 0, 0, 0, 0, 0), e_2 = (0, 1, 0, 0, 0, 0), \ldots, e_6 = (0, 0, 0, 0, 0, 1).$$

Of course, it is not enough to just assert that $\mathbb{F}^{m \times n}$ is a vector space since we have yet to verify that the above defined operations of addition and scalar multiplication satisfy the vector space axioms. The proof of the following theorem is straightforward and something that you should work through for practice with matrix notation.

Theorem A.2.3. *Given positive integers* $m, n \in \mathbb{Z}_+$ *and the operations of matrix addition and scalar multiplication as defined above, the set* $\mathbb{F}^{m \times n}$ *of all* $m \times n$ *matrices satisfies each of the following properties.*

(1) *(associativity of matrix addition) Given any three matrices* $A, B, C \in \mathbb{F}^{m \times n}$,

$$(A + B) + C = A + (B + C).$$

(2) *(additive identity for matrix addition) Given any matrix* $A \in \mathbb{F}^{m \times n}$,

$$A + 0_{m \times n} = 0_{m \times n} + A = A.$$

(3) *(additive inverses for matrix addition) Given any matrix* $A \in \mathbb{F}^{m \times n}$, *there exists a matrix* $-A \in \mathbb{F}^{m \times n}$ *such that*

$$A + (-A) = (-A) + A = 0_{m \times n}.$$

(4) *(commutativity of matrix addition) Given any two matrices* $A, B \in \mathbb{F}^{m \times n}$,

$$A + B = B + A.$$

(5) *(associativity of scalar multiplication) Given any matrix* $A \in \mathbb{F}^{m \times n}$ *and any two scalars* $\alpha, \beta \in \mathbb{F}$,

$$(\alpha \beta) A = \alpha (\beta A).$$

(6) *(multiplicative identity for scalar multiplication) Given any matrix* $A \in \mathbb{F}^{m \times n}$ *and denoting by* 1 *the multiplicative identity of* $\mathbb{F}$,

$$1A = A.$$

(7) *(distributivity of scalar multiplication) Given any two matrices* $A, B \in \mathbb{F}^{m \times n}$ *and any two scalars* $\alpha, \beta \in \mathbb{F}$,

$$(\alpha + \beta) A = \alpha A + \beta A \quad and \quad \alpha (A + B) = \alpha A + \alpha B.$$

In other words, $\mathbb{F}^{m \times n}$ *forms a vector space under the operations of matrix addition and scalar multiplication.*

As a consequence of Theorem A.2.3, every property that holds for an arbitrary vector space can be taken as a property of $\mathbb{F}^{m \times n}$ specifically. We highlight some of these properties in the following corollary to Theorem A.2.3.

Corollary A.2.4. *Given positive integers* $m, n \in \mathbb{Z}_+$ *and the operations of matrix addition and scalar multiplication as defined above, the set* $\mathbb{F}^{m \times n}$ *of all* $m \times n$ *matrices satisfies each of the following properties:*

(1) Given any matrix $A \in \mathbb{F}^{m \times n}$, given any scalar $\alpha \in \mathbb{F}$, and denoting by 0 the additive identity of $\mathbb{F}$,

$$0A = A \quad and \quad \alpha 0_{m \times n} = 0_{m \times n}.$$

(2) Given any matrix $A \in \mathbb{F}^{m \times n}$ and any scalar $\alpha \in \mathbb{F}$,

$$\alpha A = 0 \implies either \quad \alpha = 0 \quad or \quad A = 0_{m \times n}.$$

(3) Given any matrix $A \in \mathbb{F}^{m \times n}$ and any scalar $\alpha \in \mathbb{F}$,

$$-(\alpha A) = (-\alpha)A = \alpha(-A).$$

In particular, the additive inverse $-A$ of A is given by $-A = (-1)A$, where -1 denotes the additive inverse for the additive identity of $\mathbb{F}$.

While one could prove Corollary A.2.4 directly from definitions, the point of recognizing $\mathbb{F}^{m \times n}$ as a vector space is that you get to use these results without worrying about their proof. Moreover, there is no need for separate proofs for $\mathbb{F} = \mathbb{R}$ and $\mathbb{F} = \mathbb{C}$.

A.2.2 *Multiplication of matrices*

Let $r, s, t \in \mathbb{Z}_+$ be positive integers, $A = (a_{ij}) \in \mathbb{F}^{r \times s}$ be an $r \times s$ matrix, and $B = (b_{ij}) \in \mathbb{F}^{s \times t}$ be an $s \times t$ matrix. Then **matrix multiplication** $AB = ((ab)_{ij})_{r \times t}$ is defined by

$$(ab)_{ij} = \sum_{k=1}^{s} a_{ik} b_{kj}.$$

In particular, note that the "i, j entry" of the matrix product AB involves a summation over the positive integer $k = 1, \ldots, s$, where s is both the number of columns in A and the number of rows in B. Thus, this multiplication is only defined when the "middle" dimension of each matrix is the same:

$$(a_{ij})_{r \times s} (b_{ij})_{s \times t} = r \left\{ \begin{bmatrix} a_{11} & \cdots & a_{1s} \\ \vdots & \ddots & \vdots \\ a_{r1} & \cdots & a_{rs} \end{bmatrix} \underbrace{\begin{bmatrix} b_{11} & \cdots & b_{1t} \\ \vdots & \ddots & \vdots \\ b_{s1} & \cdots & b_{st} \end{bmatrix}}_{t} \right\}_{s}$$

$$= \underbrace{\begin{bmatrix} \sum_{k=1}^{s} a_{1k} b_{k1} & \cdots & \sum_{k=1}^{s} a_{1k} b_{kt} \\ \vdots & \ddots & \vdots \\ \sum_{k=1}^{s} a_{rk} b_{k1} & \cdots & \sum_{k=1}^{s} a_{rk} b_{kt} \end{bmatrix}}_{t} \Bigg\}_{r}$$

Alternatively, if we let $n \in \mathbb{Z}_+$ be a positive integer, then another way of viewing matrix multiplication is through the use of the standard inner product on $\mathbb{F}^n =$

$\mathbb{F}^{1 \times n} = \mathbb{F}^{n \times 1}$. In particular, we define the **dot product** (a.k.a. **Euclidean inner product**) of the row vector $x = (x_{1j}) \in \mathbb{F}^{1 \times n}$ and the column vector $y = (y_{i1}) \in \mathbb{F}^{n \times 1}$ to be

$$x \cdot y = \begin{bmatrix} x_{11}, & \cdots, & x_{1n} \end{bmatrix} \cdot \begin{bmatrix} y_{11} \\ \vdots \\ y_{n1} \end{bmatrix} = \sum_{k=1}^{n} x_{1k} y_{k1} \in \mathbb{F}.$$

We can then decompose matrices $A = (a_{ij})_{r \times s}$ and $B = (b_{ij})_{s \times t}$ into their constituent **row vectors** by fixing a positive integer $k \in \mathbb{Z}_+$ and setting

$$A^{(k, \cdot)} = \begin{bmatrix} a_{k1}, & \cdots, & a_{ks} \end{bmatrix} \in \mathbb{F}^{1 \times s} \quad \text{and} \quad B^{(k, \cdot)} = \begin{bmatrix} b_{k1}, & \cdots, & b_{kt} \end{bmatrix} \in \mathbb{F}^{1 \times t}.$$

Similarly, fixing $\ell \in \mathbb{Z}_+$, we can also decompose A and B into the **column vectors**

$$A^{(\cdot, \ell)} = \begin{bmatrix} a_{1\ell} \\ \vdots \\ a_{r\ell} \end{bmatrix} \in \mathbb{F}^{r \times 1} \quad \text{and} \quad B^{(\cdot, \ell)} = \begin{bmatrix} b_{1\ell} \\ \vdots \\ b_{s\ell} \end{bmatrix} \in \mathbb{F}^{s \times 1}.$$

It follows that the product AB is the following matrix of dot products:

$$AB = \begin{bmatrix} A^{(1, \cdot)} \cdot B^{(\cdot, 1)} & \cdots & A^{(1, \cdot)} \cdot B^{(\cdot, t)} \\ \vdots & \ddots & \vdots \\ A^{(r, \cdot)} \cdot B^{(\cdot, 1)} & \cdots & A^{(r, \cdot)} \cdot B^{(\cdot, t)} \end{bmatrix} \in \mathbb{F}^{r \times t}.$$

Example A.2.5. With the notation as in Example A.1.3, the reader is advised to use the above definitions to verify that the following matrix products hold.

$$AC = \begin{bmatrix} 3 \\ -1 \\ 1 \end{bmatrix} \begin{bmatrix} 1, & 4, & 2 \end{bmatrix} = \begin{bmatrix} 3 & 12 & 6 \\ -1 & -4 & -2 \\ 1 & 4 & 2 \end{bmatrix} \in \mathbb{F}^{3 \times 3},$$

$$CA = \begin{bmatrix} 1, & 4, & 2 \end{bmatrix} \begin{bmatrix} 3 \\ -1 \\ 1 \end{bmatrix} = 3 - 4 + 2 = 1 \in \mathbb{F},$$

$$B^2 = BB = \begin{bmatrix} 4 & -1 \\ 0 & 2 \end{bmatrix} \begin{bmatrix} 4 & -1 \\ 0 & 2 \end{bmatrix} = \begin{bmatrix} 16 & -6 \\ 0 & 4 \end{bmatrix} \in \mathbb{F}^{2 \times 2},$$

$$CE = \begin{bmatrix} 1, & 4, & 2 \end{bmatrix} \begin{bmatrix} 6 & 1 & 3 \\ -1 & 1 & 2 \\ 4 & 1 & 3 \end{bmatrix} = \begin{bmatrix} 10, & 7, & 17 \end{bmatrix} \in \mathbb{F}^{1 \times 3}, \text{ and}$$

$$DA = \begin{bmatrix} 1 & 5 & 2 \\ -1 & 0 & 1 \\ 3 & 2 & 4 \end{bmatrix} \begin{bmatrix} 3 \\ -1 \\ 1 \end{bmatrix} = \begin{bmatrix} 0 \\ -2 \\ 11 \end{bmatrix} \in \mathbb{F}^{3 \times 1}.$$

Note, though, that B cannot be multiplied by any of the other matrices, nor does it make sense to try to form the products AD, AE, DC, and EC due to the inherent size mismatches.

As illustrated in Example A.2.5 above, matrix multiplication is not a commutative operation (since, e.g., $AC \in \mathbb{F}^{3 \times 3}$ while $CA \in \mathbb{F}^{1 \times 1}$). Nonetheless, despite the complexity of its definition, the matrix product otherwise satisfies many familiar properties of a multiplication operation. We summarize the most basic of these properties in the following theorem.

Theorem A.2.6. *Let $r, s, t, u \in \mathbb{Z}_+$ be positive integers.*

(1) (associativity of matrix multiplication) Given $A \in \mathbb{F}^{r \times s}$, $B \in \mathbb{F}^{s \times t}$, and $C \in \mathbb{F}^{t \times u}$,

$$A(BC) = (AB)C.$$

(2) (distributivity of matrix multiplication) Given $A \in \mathbb{F}^{r \times s}$, $B, C \in \mathbb{F}^{s \times t}$, and $D \in \mathbb{F}^{t \times u}$,

$$A(B + C) = AB + AC \quad and \quad (B + C)D = BD + CD.$$

(3) (compatibility with scalar multiplication) Given $A \in \mathbb{F}^{r \times s}$, $B \in \mathbb{F}^{s \times t}$, and $\alpha \in \mathbb{F}$,

$$\alpha(AB) = (\alpha A)B = A(\alpha B).$$

Moreover, given any positive integer $n \in \mathbb{Z}_+$, $\mathbb{F}^{n \times n}$ is an algebra over $\mathbb{F}$.

As with Theorem A.2.3, you should work through a proof of each part of Theorem A.2.6 (and especially of the first part) in order to practice manipulating the indices of entries correctly. We state and prove a useful followup to Theorems A.2.3 and A.2.6 as an illustration.

Theorem A.2.7. *Let $A, B \in \mathbb{F}^{n \times n}$ be upper triangular matrices and $c \in \mathbb{F}$ be any scalar. Then each of the following properties hold:*

(1) cA is upper triangular.
(2) $A + B$ is upper triangular.
(3) AB is upper triangular.

In other words, the set of all $n \times n$ upper triangular matrices forms an algebra over $\mathbb{F}$.

Moreover, each of the above statements still holds when upper triangular *is replaced by* lower triangular.

Proof. The proofs of Parts 1 and 2 are straightforward and follow directly from the appropriate definitions. Moreover, the proof of the case for lower triangular matrices follows from the fact that a matrix A is upper triangular if and only if A^T is lower triangular, where A^T denotes the transpose of A. (See Section A.5.1 for the definition of transpose.)

To prove Part 3, we start from the definition of the matrix product. Denoting $A = (a_{ij})$ and $B = (b_{ij})$, note that $AB = ((ab)_{ij})$ is an $n \times n$ matrix having "i-j entry" given by

$$(ab)_{ij} = \sum_{k=1}^{n} a_{ik} b_{kj}.$$

Since A and B are upper triangular, we have that $a_{ik} = 0$ when $i > k$ and that $b_{kj} = 0$ when $k > j$. Thus, to obtain a non-zero summand $a_{ik} b_{kj} \neq 0$, we must have both $a_{ik} \neq 0$, which implies that $i \leq k$, and $b_{kj} \neq 0$, which implies that $k \leq j$. In particular, these two conditions are simultaneously satisfiable only when $i \leq j$. Therefore, $(ab)_{ij} = 0$ when $i > j$, from which AB is upper triangular.   □

At the same time, you should be careful not to blithely perform operations on matrices as you would with numbers. The fact that matrix multiplication is not a commutative operation should make it clear that significantly more care is required with matrix arithmetic. As another example, given a positive integer $n \in \mathbb{Z}_+$, the set $\mathbb{F}^{n \times n}$ has what are called **zero divisors**. That is, there exist non-zero matrices $A, B \in \mathbb{F}^{n \times n}$ such that $AB = 0_{n \times n}$:

$$\begin{bmatrix} 0 & 1 \\ 0 & 0 \end{bmatrix}^2 = \begin{bmatrix} 0 & 1 \\ 0 & 0 \end{bmatrix} \begin{bmatrix} 0 & 1 \\ 0 & 0 \end{bmatrix} = \begin{bmatrix} 0 & 0 \\ 0 & 0 \end{bmatrix} = 0_{2 \times 2}.$$

Moreover, note that there exist matrices $A, B, C \in \mathbb{F}^{n \times n}$ such that $AB = AC$ but $B \neq C$:

$$\begin{bmatrix} 0 & 1 \\ 0 & 0 \end{bmatrix} \begin{bmatrix} 1 & 0 \\ 0 & 0 \end{bmatrix} = 0_{2 \times 2} = \begin{bmatrix} 0 & 1 \\ 0 & 0 \end{bmatrix} \begin{bmatrix} 0 & 1 \\ 0 & 0 \end{bmatrix}.$$

As a result, we say that the set $\mathbb{F}^{n \times n}$ fails to have the so-called **cancellation property**. This failure is a direct result of the fact that there are non-zero matrices in $\mathbb{F}^{n \times n}$ that have no multiplicative inverse. We discuss matrix invertibility at length in the next section and define a special subset $GL(n, \mathbb{F}) \subset \mathbb{F}^{n \times n}$ upon which the cancellation property does hold.

A.2.3 *Invertibility of square matrices*

In this section, we characterize square matrices for which multiplicative inverses exist.

Definition A.2.8. Given a positive integer $n \in \mathbb{Z}_+$, we say that the square matrix $A \in \mathbb{F}^{n \times n}$ is **invertible** (a.k.a. **non-singular**) if there exists a square matrix $B \in \mathbb{F}^{n \times n}$ such that

$$AB = BA = I_n.$$

We use $GL(n, \mathbb{F})$ to denote the set of all invertible $n \times n$ matrices having entries from $\mathbb{F}$.

One can prove that, if the multiplicative inverse of a matrix exists, then the inverse is unique. As such, we will usually denote the so-called **inverse matrix** of $A \in GL(n, \mathbb{F})$ by A^{-1}. Note that the zero matrix $0_{n \times n} \notin GL(n, \mathbb{F})$. This means that $GL(n, \mathbb{F})$ *is not* a vector subspace of $\mathbb{F}^{n \times n}$.

Since matrix multiplication is not a commutative operation, care must be taken when working with the multiplicative inverses of invertible matrices. In particular, many of the algebraic properties for multiplicative inverses of scalars, when properly modified, continue to hold. We summarize the most basic of these properties in the following theorem.

Theorem A.2.9. *Let $n \in \mathbb{Z}_+$ be a positive integer and $A, B \in GL(n, \mathbb{F})$. Then*

(1) the inverse matrix $A^{-1} \in GL(n, \mathbb{F})$ and satisfies $(A^{-1})^{-1} = A$.
(2) the matrix power $A^m \in GL(n, \mathbb{F})$ and satisfies $(A^m)^{-1} = (A^{-1})^m$, where $m \in \mathbb{Z}_+$ is any positive integer.
(3) the matrix $\alpha A \in GL(n, \mathbb{F})$ and satisfies $(\alpha A)^{-1} = \alpha^{-1} A^{-1}$, where $\alpha \in \mathbb{F}$ is any non-zero scalar.
(4) the product $AB \in GL(n, \mathbb{F})$ and has inverse given by the formula

$$(AB)^{-1} = B^{-1} A^{-1}.$$

*Moreover, $GL(n, \mathbb{F})$ has the **cancellation property**. In other words, given any three matrices $A, B, C \in GL(n, \mathbb{F})$, if $AB = AC$, then $B = C$.*

At the same time, it is important to note that the zero matrix is not the only non-invertible matrix. As an illustration of the subtlety involved in understanding invertibility, we give the following theorem for the 2×2 case.

Theorem A.2.10. *Let $A = \begin{bmatrix} a_{11} & a_{12} \\ a_{21} & a_{22} \end{bmatrix} \in \mathbb{F}^{2 \times 2}$. Then A is invertible if and only if A satisfies*

$$a_{11}a_{22} - a_{12}a_{21} \neq 0.$$

Moreover, if A is invertible, then

$$A^{-1} = \begin{bmatrix} \dfrac{a_{22}}{a_{11}a_{22} - a_{12}a_{21}} & \dfrac{-a_{12}}{a_{11}a_{22} - a_{12}a_{21}} \\ \dfrac{-a_{21}}{a_{11}a_{22} - a_{12}a_{21}} & \dfrac{a_{11}}{a_{11}a_{22} - a_{12}a_{21}} \end{bmatrix}.$$

A more general theorem holds for larger matrices. Its statement requires the notion of determinant and we refer the reader to Chapter 8 for the definition of the determinant. For completeness, we state the result here.

Theorem A.2.11. *Let $n \in \mathbb{Z}_+$ be a positive integer, and let $A = (a_{ij}) \in \mathbb{F}^{n \times n}$ be an $n \times n$ matrix. Then A is invertible if and only if $\det(A) \neq 0$. Moreover, if A is invertible, then the "i,j entry" of A^{-1} is $A_{ji}/\det(A)$. Here, $A_{ij} = (-1)^{i+j} M_{ij}$,* and M_{ij} is the determinant of the matrix obtained when both the ith row and j^{th} column are removed from A.

We close this section by noting that the set $GL(n, \mathbb{F})$ of all invertible $n \times n$ matrices over $\mathbb{F}$ is often called the **general linear group**. This set has many important uses in mathematics and there are several equivalent notations for it, including $GL_n(\mathbb{F})$ and $GL(\mathbb{F}^n)$, and sometimes simply $GL(n)$ or GL_n if it is not important to emphasize the dependence on $\mathbb{F}$. Note that the usage of the term "group" in the name "general linear group" has a technical meaning: $GL(n, \mathbb{F})$ forms a group under matrix multiplication, which is non-abelian if $n \geq 2$. (See Section C.2 for the definition of a group.)

A.3 Solving linear systems by factoring the coefficient matrix

There are many ways in which one might try to solve a given system of linear equations. This section is primarily devoted to describing two particularly popular techniques, both of which involve factoring the coefficient matrix for the system into a product of simpler matrices. These techniques are also at the heart of many frequently used numerical (i.e., computer-assisted) applications of Linear Algebra.

Note that the factorization of complicated objects into simpler components is an extremely common problem solving technique in mathematics. E.g., we will often factor a given polynomial into several polynomials of lower degree, and one can similarly use the prime factorization for an integer in order to simplify certain numerical computations.

A.3.1 *Factorizing matrices using Gaussian elimination*

In this section, we discuss a particularly significant factorization for matrices known as **Gaussian elimination** (a.k.a. **Gauss-Jordan elimination**). Gaussian elimination can be used to express any matrix as a product involving one matrix in so-called **reduced row-echelon form** and one or more so-called **elementary matrices**. Then, once such a factorization has been found, we can immediately solve any linear system that has the factorized matrix as its coefficient matrix. Moreover, the underlying technique for arriving at such a factorization is essentially an extension of the techniques already familiar to you for solving small systems of linear equations by hand.

Let $m, n \in \mathbb{Z}_+$ denote positive integers, and suppose that $A \in \mathbb{F}^{m \times n}$ is an $m \times n$ matrix over $\mathbb{F}$. Then, following Section A.2.2, we will make extensive use of $A^{(i, \cdot)}$ and $A^{(\cdot, j)}$ to denote the row vectors and column vectors of A, respectively.

Definition A.3.1. Let $A \in \mathbb{F}^{m \times n}$ be an $m \times n$ matrix over $\mathbb{F}$. Then we say that A is in **row-echelon form** (abbreviated **REF**) if the rows of A satisfy the following conditions:

(1) either $A^{(1,\cdot)}$ is the zero vector or the first non-zero entry in $A^{(1,\cdot)}$ (when read from left to right) is a one.
(2) for $i = 1, \ldots, m$, if any row vector $A^{(i,\cdot)}$ is the zero vector, then each subsequent row vector $A^{(i+1,\cdot)}, \ldots, A^{(m,\cdot)}$ is also the zero vector.
(3) for $i = 2, \ldots, m$, if some $A^{(i,\cdot)}$ is not the zero vector, then the first non-zero entry (when read from left to right) is a one and occurs to the right of the initial one in $A^{(i-1,\cdot)}$.

The initial leading one in each non-zero row is called a **pivot**. We furthermore say that A is in **reduced row-echelon form** (abbreviated **RREF**) if

(4) for each column vector $A^{(\cdot,j)}$ containing a pivot ($j = 2, \ldots, n$), the pivot is the only non-zero element in $A^{(\cdot,j)}$.

The motivation behind Definition A.3.1 is that matrix equations having their coefficient matrix in RREF (and, in some sense, also REF) are particularly easy to solve. Note, in particular, that the only square matrix in RREF without zero rows is the identity matrix.

Example A.3.2. The following matrices are all in REF:

$$A_1 = \begin{bmatrix} 1 & 1 & 1 & 1 \\ 0 & 1 & 1 & 1 \\ 0 & 0 & 1 & 1 \end{bmatrix}, \ A_2 = \begin{bmatrix} 1 & 1 & 1 & 0 \\ 0 & 1 & 1 & 0 \\ 0 & 0 & 1 & 0 \end{bmatrix}, \ A_3 = \begin{bmatrix} 1 & 1 & 0 & 1 \\ 0 & 1 & 1 & 0 \\ 0 & 0 & 0 & 1 \end{bmatrix}, \ A_4 = \begin{bmatrix} 1 & 1 & 0 & 0 \\ 0 & 0 & 1 & 0 \\ 0 & 0 & 0 & 1 \end{bmatrix},$$

$$A_5 = \begin{bmatrix} 1 & 0 & 1 & 0 \\ 0 & 0 & 0 & 1 \\ 0 & 0 & 0 & 0 \end{bmatrix}, \ A_6 = \begin{bmatrix} 0 & 0 & 1 & 0 \\ 0 & 0 & 0 & 1 \\ 0 & 0 & 0 & 0 \end{bmatrix}, \ A_7 = \begin{bmatrix} 0 & 0 & 0 & 1 \\ 0 & 0 & 0 & 0 \\ 0 & 0 & 0 & 0 \end{bmatrix}, \ A_8 = \begin{bmatrix} 0 & 0 & 0 & 0 \\ 0 & 0 & 0 & 0 \\ 0 & 0 & 0 & 0 \end{bmatrix}.$$

However, only A_8 through A_8 are in RREF, as you should verify. Moreover, if we take the transpose of each of these matrices (as defined in Section A.5.1), then only A_6^T, A_7^T, and A_8^T are in RREF.

Example A.3.3.

(1) Consider the following matrix in RREF:

$$A = \begin{bmatrix} 1 & 0 & 0 & 0 \\ 0 & 1 & 0 & 0 \\ 0 & 0 & 1 & 0 \\ 0 & 0 & 0 & 1 \end{bmatrix}.$$

Given any vector $b = (b_i) \in \mathbb{F}^4$, the matrix equation $Ax = b$ corresponds to the system of equations

$$\left. \begin{aligned} x_1 \quad\quad\quad\quad &= b_1 \\ x_2 \quad\quad\quad &= b_2 \\ x_3 \quad\quad &= b_3 \\ x_4 &= b_4 \end{aligned} \right\}.$$

Since A is in RREF (in fact, $A = I_4$ is the 4×4 identity matrix), we can immediately conclude that the matrix equation $Ax = b$ has the solution $x = b$ for any choice of b.

Moreover, as we will see in Section A.4.2, $x = b$ is the only solution to this system.

(2) Consider the following matrix in RREF:

$$A = \begin{bmatrix} 1 & 6 & 0 & 0 & 4 & -2 \\ 0 & 0 & 1 & 0 & 3 & 1 \\ 0 & 0 & 0 & 1 & 5 & 2 \\ 0 & 0 & 0 & 0 & 0 & 0 \end{bmatrix}.$$

Given any vector $b = (b_i) \in \mathbb{F}^4$, the matrix equation $Ax = b$ corresponds to the system of equations

$$\left. \begin{array}{l} x_1 + 6x_2 \quad\quad\quad + 4x_5 - 2x_6 = b_1 \\ \quad\quad\quad x_3 \quad + 3x_5 + \quad x_6 = b_2 \\ \quad\quad\quad\quad x_4 + 5x_5 + 2x_6 = b_3 \\ \quad\quad\quad\quad\quad\quad\quad\quad\quad 0 = b_4 \end{array} \right\}.$$

Since A is in RREF, we can immediately conclude a number of facts about solutions to this system. First of all, solutions exist if and only if $b_4 = 0$. Moreover, by "solving for the pivots", we see that the system reduces to

$$\left. \begin{array}{l} x_1 = b_1 - 6x_2 - 4x_5 + 2x_6 \\ x_3 = b_2 \quad\quad - 3x_5 - \quad x_6 \\ x_4 = b_3 \quad\quad - 5x_5 - 2x_6 \end{array} \right\},$$

and so there is only enough information to specify values for x_1, x_3, and x_4 in terms of the otherwise arbitrary values for x_2, x_5, and x_6.

In this context, x_1, x_3, and x_4 are called **leading variables** since these are the variables corresponding to the pivots in A. We similarly call x_2, x_5, and x_6 **free variables** since the leading variables have been expressed in terms of these remaining variables. In particular, given any scalars $\alpha, \beta, \gamma \in \mathbb{F}$, it follows that the vector

$$x = \begin{bmatrix} x_1 \\ x_2 \\ x_3 \\ x_4 \\ x_5 \\ x_6 \end{bmatrix} = \begin{bmatrix} b_1 - 6\alpha - 4\beta + 2\gamma \\ \alpha \\ b_2 - 3\beta - \gamma \\ b_3 - 5\beta - 2\gamma \\ \beta \\ \gamma \end{bmatrix} = \begin{bmatrix} b_1 \\ 0 \\ b_2 \\ b_3 \\ 0 \\ 0 \end{bmatrix} + \begin{bmatrix} -6\alpha \\ \alpha \\ 0 \\ 0 \\ 0 \\ 0 \end{bmatrix} + \begin{bmatrix} -4\beta \\ 0 \\ -3\beta \\ -5\beta \\ \beta \\ 0 \end{bmatrix} + \begin{bmatrix} 2\gamma \\ 0 \\ -\gamma \\ -2\gamma \\ 0 \\ \gamma \end{bmatrix}$$

must satisfy the matrix equation $Ax = b$. One can also verify that every solution to the matrix equation must be of this form. It then follows that the set of all solutions should somehow be "three dimensional".

As the above examples illustrate, a matrix equation having coefficient matrix in RREF corresponds to a system of equations that can be solved with only a small amount of computation. Somewhat amazingly, any matrix can be factored into a product that involves exactly one matrix in RREF and one or more of the matrices defined as follows.

Definition A.3.4. A square matrix $E \in \mathbb{F}^{m \times m}$ is called an **elementary matrix** if it has one of the following forms:

(1) (row exchange, a.k.a. "row swap", matrix) E is obtained from the identity matrix I_m by interchanging the row vectors $I_m^{(r, \cdot)}$ and $I_m^{(s, \cdot)}$ for some particular choice of positive integers $r, s \in \{1, 2, \ldots, m\}$. I.e., in the case that $r < s$,

$$
E = \begin{bmatrix}
1 & 0 & 0 & \cdots & 0 & 0 & 0 & \cdots & 0 & 0 & 0 & \cdots & 0 \\
0 & 1 & 0 & \cdots & 0 & 0 & 0 & \cdots & 0 & 0 & 0 & \cdots & 0 \\
0 & 0 & 1 & \cdots & 0 & 0 & 0 & \cdots & 0 & 0 & 0 & \cdots & 0 \\
\vdots & \vdots & \vdots & \ddots & \vdots & \vdots & \vdots & \ddots & \vdots & \vdots & \vdots & \ddots & \vdots \\
0 & 0 & 0 & \cdots & 1 & 0 & 0 & \cdots & 0 & 0 & 0 & \cdots & 0 \\
0 & 0 & 0 & \cdots & 0 & 0 & 0 & \cdots & 0 & 1 & 0 & \cdots & 0 \\
0 & 0 & 0 & \cdots & 0 & 0 & 1 & \cdots & 0 & 0 & 0 & \cdots & 0 \\
\vdots & \vdots & \vdots & \ddots & \vdots & \vdots & \vdots & \ddots & \vdots & \vdots & \vdots & \ddots & \vdots \\
0 & 0 & 0 & \cdots & 0 & 0 & 0 & \cdots & 1 & 0 & 0 & \cdots & 0 \\
0 & 0 & 0 & \cdots & 0 & 1 & 0 & \cdots & 0 & 0 & 0 & \cdots & 0 \\
0 & 0 & 0 & \cdots & 0 & 0 & 0 & \cdots & 0 & 0 & 1 & \cdots & 0 \\
\vdots & \vdots & \vdots & \ddots & \vdots & \vdots & \vdots & \ddots & \vdots & \vdots & \vdots & \ddots & \vdots \\
0 & 0 & 0 & \cdots & 0 & 0 & 0 & \cdots & 0 & 0 & 0 & \cdots & 1
\end{bmatrix}
\begin{matrix} \\ \\ \\ \\ \\ \leftarrow r^{\text{th}} \text{ row} \\ \\ \\ \\ \leftarrow s^{\text{th}} \text{ row.} \\ \\ \\ \\ \end{matrix}
$$

(2) (row scaling matrix) E is obtained from the identity matrix I_m by replacing the row vector $I_m^{(r, \cdot)}$ with $\alpha I_m^{(r, \cdot)}$ for some choice of non-zero scalar $0 \neq \alpha \in \mathbb{F}$ and some choice of positive integer $r \in \{1, 2, \ldots, m\}$. I.e.,

$$
E = I_m + (\alpha - 1)E_{rr} = \begin{bmatrix}
1 & 0 & \cdots & 0 & 0 & 0 & \cdots & 0 \\
0 & 1 & \cdots & 0 & 0 & 0 & \cdots & 0 \\
\vdots & \vdots & \ddots & \vdots & \vdots & \vdots & \ddots & \vdots \\
0 & 0 & \cdots & 1 & 0 & 0 & \cdots & 0 \\
0 & 0 & \cdots & 0 & \alpha & 0 & \cdots & 0 \\
0 & 0 & \cdots & 0 & 0 & 1 & \cdots & 0 \\
\vdots & \vdots & \ddots & \vdots & \vdots & \vdots & \ddots & \vdots \\
0 & 0 & \cdots & 0 & 0 & 0 & \cdots & 1
\end{bmatrix}
\begin{matrix} \\ \\ \\ \\ \leftarrow r^{\text{th}} \text{ row,} \\ \\ \\ \end{matrix}
$$

where E_{rr} is the matrix having "r, r entry" equal to one and all other entries equal to zero. (Recall that E_{rr} was defined in Section A.2.1 as a standard basis vector for the vector space $\mathbb{F}^{m \times m}$.)

(3) (row combination, a.k.a. "row sum", matrix) E is obtained from the identity matrix I_m by replacing the row vector $I_m^{(r,\cdot)}$ with $I_m^{(r,\cdot)} + \alpha I_m^{(s,\cdot)}$ for some choice of scalar $\alpha \in \mathbb{F}$ and some choice of positive integers $r, s \in \{1, 2, \ldots, m\}$. I.e., in the case that $r < s$,

$$E = I_m + \alpha E_{rs} = \begin{bmatrix} 1\,0\,0\,\cdots\,0\,0\,0\,\cdots\,0\,0\,0\,\cdots\,0 \\ 0\,1\,0\,\cdots\,0\,0\,0\,\cdots\,0\,0\,0\,\cdots\,0 \\ 0\,0\,1\,\cdots\,0\,0\,0\,\cdots\,0\,0\,0\,\cdots\,0 \\ \vdots\,\vdots\,\vdots\,\ddots\,\vdots\,\vdots\,\vdots\,\ddots\,\vdots\,\vdots\,\vdots\,\ddots\,\vdots \\ 0\,0\,0\,\cdots\,1\,0\,0\,\cdots\,0\,0\,0\,\cdots\,0 \\ 0\,0\,0\,\cdots\,0\,1\,0\,\cdots\,0\,\alpha\,0\,\cdots\,0 \\ 0\,0\,0\,\cdots\,0\,0\,1\,\cdots\,0\,0\,0\,\cdots\,0 \\ \vdots\,\vdots\,\vdots\,\ddots\,\vdots\,\vdots\,\vdots\,\ddots\,\vdots\,\vdots\,\vdots\,\ddots\,\vdots \\ 0\,0\,0\,\cdots\,0\,0\,0\,\cdots\,1\,0\,0\,\cdots\,0 \\ 0\,0\,0\,\cdots\,0\,0\,0\,\cdots\,0\,1\,0\,\cdots\,0 \\ 0\,0\,0\,\cdots\,0\,0\,0\,\cdots\,0\,0\,1\,\cdots\,0 \\ \vdots\,\vdots\,\vdots\,\ddots\,\vdots\,\vdots\,\vdots\,\ddots\,\vdots\,\vdots\,\vdots\,\ddots\,\vdots \\ 0\,0\,0\,\cdots\,0\,0\,0\,\cdots\,0\,0\,0\,\cdots\,1 \end{bmatrix} \begin{matrix} \\ \\ \\ \\ \\ \leftarrow r^{\text{th}} \text{ row} \\ \\ \\ \\ \\ \\ \\ \\ \end{matrix}$$

$$\uparrow$$
$$s^{\text{th}} \text{ column}$$

where E_{rs} is the matrix having "r, s entry" equal to one and all other entries equal to zero. (E_{rs} was also defined in Section A.2.1 as a standard basis vector for $\mathbb{F}^{m \times m}$.)

The "elementary" in the name "elementary matrix" comes from the correspondence between these matrices and so-called "elementary operations" on systems of equations. In particular, each of the elementary matrices is clearly invertible (in the sense defined in Section A.2.3), just as each "elementary operation" is itself completely reversible. We illustrate this correspondence in the following example.

Example A.3.5. Define A, x, and b by

$$A = \begin{bmatrix} 2 & 5 & 3 \\ 1 & 2 & 3 \\ 1 & 0 & 8 \end{bmatrix}, \quad x = \begin{bmatrix} x_1 \\ x_2 \\ x_3 \end{bmatrix}, \quad \text{and} \quad b = \begin{bmatrix} 4 \\ 5 \\ 9 \end{bmatrix}.$$

We illustrate the correspondence between elementary matrices and "elementary" operations on the system of linear equations corresponding to the matrix equation $Ax = b$, as follows.

System of Equations	Corresponding Matrix Equation

$$\begin{aligned} 2x_1 + 5x_2 + 3x_3 &= 5 \\ x_1 + 2x_2 + 3x_3 &= 4 \\ x_1 \quad\quad + 8x_3 &= 9 \end{aligned} \qquad\qquad Ax = b$$

To begin solving this system, one might want to either multiply the first equation through by $1/2$ or interchange the first equation with one of the other equations. From a computational perspective, it is preferable to perform an interchange since multiplying through by $1/2$ would unnecessarily introduce fractions. Thus, we choose to interchange the first and second equation in order to obtain

<table>
<tr><td>System of Equations</td><td>Corresponding Matrix Equation</td></tr>
</table>

$$
\begin{aligned}
x_1 + 2x_2 + 3x_3 &= 4 \\
2x_1 + 5x_2 + 3x_3 &= 5 \\
x_1 \qquad\quad + 8x_3 &= 9
\end{aligned}
\qquad
E_0 A x = E_0 b, \text{ where } E_0 = \begin{bmatrix} 0 & 1 & 0 \\ 1 & 0 & 0 \\ 0 & 0 & 1 \end{bmatrix}.
$$

Another reason for performing the above interchange is that it now allows us to use more convenient "row combination" operations when eliminating the variable x_1 from all but one of the equations. In particular, we can multiply the first equation through by -2 and add it to the second equation in order to obtain

<table>
<tr><td>System of Equations</td><td>Corresponding Matrix Equation</td></tr>
</table>

$$
\begin{aligned}
x_1 + 2x_2 + 3x_3 &= 4 \\
x_2 - 3x_3 &= -3 \\
x_1 \qquad\quad + 8x_3 &= 9
\end{aligned}
\qquad
E_1 E_0 A x = E_1 E_0 b, \text{ where } E_1 = \begin{bmatrix} 1 & 0 & 0 \\ -2 & 1 & 0 \\ 0 & 0 & 1 \end{bmatrix}.
$$

Similarly, in order to eliminate the variable x_1 from the third equation, we can next multiply the first equation through by -1 and add it to the third equation in order to obtain

<table>
<tr><td>System of Equations</td><td>Corresponding Matrix Equation</td></tr>
</table>

$$
\begin{aligned}
x_1 + \quad 2x_2 + 3x_3 &= 4 \\
x_2 - 3x_3 &= -3 \\
-2x_2 + 5x_3 &= 5
\end{aligned}
\qquad
E_2 E_1 E_0 A x = E_2 E_1 E_0 b, \text{ where } E_2 = \begin{bmatrix} 1 & 0 & 0 \\ 0 & 1 & 0 \\ -1 & 0 & 1 \end{bmatrix}.
$$

Now that the variable x_1 only appears in the first equation, we can somewhat similarly isolate the variable x_2 by multiplying the second equation through by 2 and adding it to the third equation in order to obtain

<table>
<tr><td>System of Equations</td><td>Corresponding Matrix Equation</td></tr>
</table>

$$
\begin{aligned}
x_1 + 2x_2 + \quad 3x_3 &= 4 \\
x_2 - \quad 3x_3 &= -3 \\
-x_3 &= -1
\end{aligned}
\qquad
E_3 \cdots E_0 A x = E_3 \cdots E_0 b, \text{ where } E_3 = \begin{bmatrix} 1 & 0 & 0 \\ 0 & 1 & 0 \\ 0 & 2 & 1 \end{bmatrix}.
$$

Finally, in order to complete the process of transforming the coefficient matrix into REF, we need only rescale row three by -1. This corresponds to multiplying the third equation through by -1 in order to obtain

System of Equations   Corresponding Matrix Equation

$$x_1 + 2x_2 + 3x_3 = 4$$
$$x_2 - 3x_3 = -3$$
$$x_3 = 1$$

$E_4 \cdots E_0 A x = E_4 \cdots E_0 b$, where $E_4 = \begin{bmatrix} 1 & 0 & 0 \\ 0 & 1 & 0 \\ 0 & 0 & -1 \end{bmatrix}$.

Now that the coefficient matrix is in REF, we can already solve for the variables x_1, x_2, and x_3 using a process called **back substitution**. In other words, starting from the third equation we see that $x_3 = 1$. Using this value and solving for x_2 in the second equation, it then follows that

$$x_2 = -3 + 3x_3 = -3 + 3 = 0.$$

Similarly, by solving the first equation for x_1, it follows that

$$x_1 = 4 - 2x_2 - 3x_3 = 4 - 3 = 1.$$

From a computational perspective, this process of back substitution can be applied to solve any system of equations when the coefficient matrix of the corresponding matrix equation is in REF. However, from an algorithmic perspective, it is often more useful to continue "row reducing" the coefficient matrix in order to produce a coefficient matrix in full RREF.

There is more than one way to reach the RREF form. We choose to now work "from bottom up, and from right to left". In other words, we now multiply the third equation through by 3 and then add it to the second equation in order to obtain

System of Equations   Corresponding Matrix Equation

$$x_1 + 2x_2 + 3x_3 = 4$$
$$x_2 = 0$$
$$x_3 = 1$$

$E_5 \cdots E_0 A x = E_5 \cdots E_0 b$, where $E_5 = \begin{bmatrix} 1 & 0 & 0 \\ 0 & 1 & 3 \\ 0 & 0 & 1 \end{bmatrix}$.

Next, we can multiply the third equation through by -3 and add it to the first equation in order to obtain

System of Equations   Corresponding Matrix Equation

$$x_1 + 2x_2 = 1$$
$$x_2 = 0$$
$$x_3 = 1$$

$E_6 \cdots E_0 A x = E_6 \cdots E_0 b$, where $E_6 = \begin{bmatrix} 1 & 0 & -3 \\ 0 & 1 & 0 \\ 0 & 0 & 1 \end{bmatrix}$.

Finally, we can multiply the second equation through by -2 and add it to the first equation in order to obtain

System of Equations   Corresponding Matrix Equation

$$x_1 = 1$$
$$x_2 = 0$$
$$x_3 = 1$$

$E_7 \cdots E_0 A x = E_7 \cdots E_0 b$, where $E_7 = \begin{bmatrix} 1 & -2 & 0 \\ 0 & 1 & 0 \\ 0 & 0 & 1 \end{bmatrix}$.

Previously, we obtained a solution by using back substitution on the linear system

$$E_4 \cdots E_0 A x = E_4 \cdots E_0 b.$$

However, in many applications, it is not enough to merely find *a solution*. Instead, it is important to describe *every solution*. As we will see in the remaining sections of these notes, Linear Algebra is a very useful tool to solve this problem. In particular, we will use the theory of vector spaces and linear maps.

To close this section, we take a closer look at the following expression obtained from the above analysis:

$$E_7 E_6 \cdots E_1 E_0 A = I_3.$$

It follows from their definition that elementary matrices are invertible. In particular, each of the matrices $E_0, E_1, \ldots, E_7$ is invertible. Thus, we can use Theorem A.2.9 in order to "solve" for A:

$$A = (E_7 E_6 \cdots E_1 E_0)^{-1} I_3 = E_0^{-1} E_1^{-1} \cdots E_7^{-1} I_3.$$

In effect, since the inverse of an elementary matrix is itself easily seen to be an elementary matrix, this has factored A into the product of eight elementary matrices (namely, $E_0^{-1}, E_1^{-1}, \ldots, E_7^{-1}$) and one matrix in RREF (namely, I_3). Moreover, because each elementary matrix is invertible, we can conclude that x solves $Ax = b$ if and only if x solves

$$(E_7 E_6 \cdots E_1 E_0 A)\, x = (I_3)\, x = (E_7 E_6 \cdots E_1 E_0)\, b.$$

Consequently, given any linear system, one can use Gaussian elimination in order to reduce the problem to solving a linear system whose coefficient matrix is in RREF.

Similarly, we can conclude that the inverse of A is

$$A^{-1} = E_7 E_6 \cdots E_1 E_0 = \begin{bmatrix} 13 & -5 & -3 \\ -40 & 16 & 9 \\ 5 & -2 & 1 \end{bmatrix}.$$

Having computed this product, one could essentially "reuse" much of the above computation in order to solve the matrix equation $Ax = b'$ for several different right-hand sides $b' \in \mathbb{F}^3$. The process of "resolving" a linear system is a common practice in applied mathematics.

A.3.2 *Solving homogeneous linear systems*

In this section, we study the solutions for an important special class of linear systems. As we will see in the next section, though, solving any linear system is fundamentally dependent upon knowing how to solve these so-called **homogeneous systems**.

As usual, we use $m, n \in \mathbb{Z}_+$ to denote arbitrary positive integers.

Definition A.3.6. The system of linear equations, System (A.1), is called a **homogeneous system** if the right-hand side of each equation is zero. In other words, a homogeneous system corresponds to a matrix equation of the form

$$Ax = 0,$$

where $A \in \mathbb{F}^{m \times n}$ is an $m \times n$ matrix and x is an n-tuple of unknowns. We also call the set

$$N = \{v \in \mathbb{F}^n \mid Av = 0\}$$

the **solution space** for the homogeneous system corresponding to $Ax = 0$.

When describing the solution space for a homogeneous linear system, there are three important cases to keep in mind:

Definition A.3.7. The system of linear equations System (A.1) is called

(1) **overdetermined** if $m > n$.
(2) **square** if $m = n$.
(3) **underdetermined** if $m < n$.

In particular, we can say a great deal about underdetermined homogeneous systems, which we state as a corollary to the following more general result.

Theorem A.3.8. *Let N be the solution space for the homogeneous linear system corresponding to the matrix equation $Ax = 0$, where $A \in \mathbb{F}^{m \times n}$. Then*

(1) the zero vector $0 \in N$.
(2) N is a subspace of the vector space $\mathbb{F}^n$.

This is an amazing theorem. Since N is a subspace of $\mathbb{F}^n$, we know that either N will contain exactly one element (namely, the zero vector) or N will contain infinitely many elements.

Corollary A.3.9. *Every homogeneous system of linear equations is solved by the zero vector. Moreover, every underdetermined homogeneous system has infinitely many solutions.*

We call the zero vector the **trivial solution** for a homogeneous linear system. The fact that every homogeneous linear system has the trivial solution thus reduces solving such a system to determining if solutions other than the trivial solution exist.

One method for finding the solution space of a homogeneous system is to first use Gaussian elimination (as demonstrated in Example A.3.5) in order to factor the coefficient matrix of the system. Then, because the original linear system is homogeneous, the homogeneous system corresponding to the resulting RREF matrix will have the same solutions as the original system. In other words, if a given matrix A satisfies

$$E_k E_{k-1} \cdots E_0 A = A_0,$$

where each E_i is an elementary matrix and A_0 is an RREF matrix, then the matrix equation $Ax = 0$ has the exact same solution set as $A_0 x = 0$ since $E_0^{-1} E_1^{-1} \cdots E_k^{-1} 0 = 0$.

Example A.3.10. In the following examples, we illustrate the process of determining the solution space for a homogeneous linear system having coefficient matrix in RREF.

(1) Consider the matrix equation $Ax = 0$, where A is the matrix given by

$$A = \begin{bmatrix} 1 & 0 & 0 \\ 0 & 1 & 0 \\ 0 & 0 & 1 \\ 0 & 0 & 0 \end{bmatrix}.$$

This corresponds to an overdetermined homogeneous system of linear equations. Moreover, since there are no free variables (as defined in Example A.3.3), it should be clear that this system has only the trivial solution. Thus, $N = \{0\}$.

(2) Consider the matrix equation $Ax = 0$, where A is the matrix given by

$$A = \begin{bmatrix} 1 & 0 & 1 \\ 0 & 1 & 1 \\ 0 & 0 & 0 \\ 0 & 0 & 0 \end{bmatrix}.$$

This corresponds to an overdetermined homogeneous system of linear equations. Unlike the above example, we see that x_3 is a free variable for this system, and so we would expect the solution space to contain more than just the zero vector. As in Example A.3.3, we can solve for the leading variables in terms of the free variable in order to obtain

$$\left. \begin{aligned} x_1 &= -x_3 \\ x_2 &= -x_3 \end{aligned} \right\}.$$

It follows that, given any scalar $\alpha \in \mathbb{F}$, every vector of the form

$$x = \begin{bmatrix} x_1 \\ x_2 \\ x_3 \end{bmatrix} = \begin{bmatrix} -\alpha \\ -\alpha \\ \alpha \end{bmatrix} = \alpha \begin{bmatrix} -1 \\ -1 \\ 1 \end{bmatrix}$$

is a solution to $Ax = 0$. Therefore,

$$N = \left\{ (x_1, x_2, x_3) \in \mathbb{F}^3 \mid x_1 = -x_3, x_2 = -x_3 \right\} = \text{span}\left((-1, -1, 1)\right).$$

(3) Consider the matrix equation $Ax = 0$, where A is the matrix given by

$$A = \begin{bmatrix} 1 & 1 & 1 \\ 0 & 0 & 0 \\ 0 & 0 & 0 \end{bmatrix}.$$

This corresponds to a square homogeneous system of linear equations with two free variables. Thus, using the same technique as in the previous example, we can solve for the leading variable in order to obtain $x_1 = -x_2 - x_3$. It follows that, given any scalars $\alpha, \beta \in \mathbb{F}$, every vector of the form

$$x = \begin{bmatrix} x_1 \\ x_2 \\ x_3 \end{bmatrix} = \begin{bmatrix} -\alpha - \beta \\ \alpha \\ \beta \end{bmatrix} = \alpha \begin{bmatrix} -1 \\ 1 \\ 0 \end{bmatrix} + \beta \begin{bmatrix} -1 \\ 0 \\ 1 \end{bmatrix}$$

is a solution to $Ax = 0$. Therefore,

$$N = \left\{ (x_1, x_2, x_3) \in \mathbb{F}^3 \mid x_1 + x_2 + x_3 = 0 \right\} = \text{span} \left((-1, 1, 0), (-1, 0, 1) \right).$$

A.3.3   *Solving inhomogeneous linear systems*

In this section, we demonstrate the relationship between arbitrary linear systems and homogeneous linear systems. Specifically, we will see that it takes little more work to solve a general linear system than it does to solve the homogeneous system associated to it.

As usual, we use $m, n \in \mathbb{Z}_+$ to denote arbitrary positive integers.

Definition A.3.11. The system of linear equations System (A.1) is called an **inhomogeneous system** if the right-hand side of at least one equation is not zero. In other words, an inhomogeneous system corresponds to a matrix equation of the form

$$Ax = b,$$

where $A \in \mathbb{F}^{m \times n}$ is an $m \times n$ matrix, x is an n-tuple of unknowns, and $b \in \mathbb{F}^m$ is a vector having at least one non-zero component. We also call the set

$$U = \{v \in \mathbb{F}^n \mid Av = b\}$$

the **solution set** for the linear system corresponding to $Ax = b$.

As illustrated in Example A.3.3, the zero vector cannot be a solution for an inhomogeneous system. Consequently, the solution set U for an inhomogeneous linear system will never be a subspace of any vector space. Instead, it will be a related algebraic structure as described in the following theorem.

Theorem A.3.12. *Let U be the solution space for the inhomogeneous linear system corresponding to the matrix equation $Ax = b$, where $A \in \mathbb{F}^{m \times n}$ and $b \in \mathbb{F}^m$ is a vector having at least one non-zero component. Then, given any element $u \in U$, we have that*

$$U = u + N = \{u + n \mid n \in N\},$$

where N is the solution space to $Ax = 0$, or the kernel of A. In other words, if $B = (n^{(1)}, n^{(2)}, \ldots, n^{(k)})$ is a list of vectors forming a basis for N, then every element of U can be written in the form

$$u + \alpha_1 n^{(1)} + \alpha_2 n^{(2)} + \ldots + \alpha_k n^{(k)}$$

for some choice of scalars $\alpha_1, \alpha_2, \ldots, \alpha_k \in \mathbb{F}$.

As a consequence of this theorem, we can conclude that inhomogeneous linear systems behave a lot like homogeneous systems. The main difference is that inhomogeneous systems are not necessarily solvable. This, then, creates three possibilities: an inhomogeneous linear system will either have no solution, a unique solution, or infinitely many solutions. An important special case is as follows.

Corollary A.3.13. *Every overdetermined inhomogeneous linear system will necessarily be unsolvable for some choice of values for the right-hand sides of the equations.*

The solution set U for an inhomogeneous linear system is called an **affine subspace** of $\mathbb{F}^n$ since it is a genuine subspace of $\mathbb{F}^n$ that has been "offset" by a vector $u \in \mathbb{F}^n$. Any set having this structure might also be called a **coset** (when used in the context of Group Theory) or a **linear manifold** (when used in a geometric context such as a discussion of hyperplanes).

In order to actually find the solution set for an inhomogeneous linear system, we rely on Theorem A.3.12. Given an $m \times n$ matrix $A \in \mathbb{F}^{m \times n}$ and a non-zero vector $b \in \mathbb{F}^m$, we call $Ax = 0$ the **associated homogeneous matrix equation** to the inhomogeneous matrix equation $Ax = b$. Then, according to Theorem A.3.12, U can be found by first finding the solution space N for the associated equation $Ax = 0$ and then finding any so-called **particular solution** $u \in \mathbb{F}^n$ to $Ax = b$.

As with homogeneous systems, one can first use Gaussian elimination in order to factorize A, and so we restrict the following examples to the special case of RREF matrices.

Example A.3.14. The following examples use the same matrices as in Example A.3.10.

(1) Consider the matrix equation $Ax = b$, where A is the matrix given by

$$A = \begin{bmatrix} 1 & 0 & 0 \\ 0 & 1 & 0 \\ 0 & 0 & 1 \\ 0 & 0 & 0 \end{bmatrix}$$

and $b \in \mathbb{F}^4$ has at least one non-zero component. Then $Ax = b$ corresponds to an overdetermined inhomogeneous system of linear equations and will not necessarily be solvable for all possible choices of b.

In particular, note that the bottom row $A^{(4,\cdot)}$ of A corresponds to the equation

$$0 = b_4,$$

from which we conclude that $Ax = b$ has no solution unless the fourth component of b is zero. Furthermore, the remaining rows of A correspond to the equations

$$x_1 = b_1, \; x_2 = b_2, \text{ and } x_3 = b_3.$$

It follows that, given any vector $b \in \mathbb{F}^n$ with fourth component zero, $x = b$ is the only solution to $Ax = b$. In other words, $U = \{b\}$.

(2) Consider the matrix equation $Ax = b$, where A is the matrix given by

$$A = \begin{bmatrix} 1 & 0 & 1 \\ 0 & 1 & 1 \\ 0 & 0 & 0 \\ 0 & 0 & 0 \end{bmatrix}$$

and $b \in \mathbb{F}^4$. This corresponds to an overdetermined inhomogeneous system of linear equations. Note, in particular, that the bottom two rows of the matrix correspond to the equations $0 = b_3$ and $0 = b_4$, from which we see that $Ax = b$ has no solution unless the third and fourth components of the vector b are both zero. Furthermore, we conclude from the remaining rows of the matrix that x_3 is a free variable for this system and that

$$\left. \begin{array}{l} x_1 = b_1 - x_3 \\ x_2 = b_2 - x_3 \end{array} \right\} .$$

It follows that, given any scalar $\alpha \in \mathbb{F}$, every vector of the form

$$x = \begin{bmatrix} x_1 \\ x_2 \\ x_3 \end{bmatrix} = \begin{bmatrix} b_1 - \alpha \\ b_2 - \alpha \\ \alpha \end{bmatrix} = \begin{bmatrix} b_1 \\ b_2 \\ 0 \end{bmatrix} + \alpha \begin{bmatrix} -1 \\ -1 \\ 1 \end{bmatrix} = u + \alpha n$$

is a solution to $Ax = b$. Recall from Example A.3.10 that the solution space for the associated homogeneous matrix equation $Ax = 0$ is

$$N = \left\{ (x_1, x_2, x_3) \in \mathbb{F}^3 \mid x_1 = -x_3, x_2 = -x_3 \right\} = \operatorname{span}\left((-1, -1, 1) \right).$$

Thus, in the language of Theorem A.3.12, we have that u is a particular solution for $Ax = b$ and that (n) is a basis for N. Therefore, the solution set for $Ax = b$ is

$$U = (b_1, b_2, 0) + N = \left\{ (x_1, x_2, x_3) \in \mathbb{F}^3 \mid x_1 = b_1 - x_3, x_2 = b_2 - x_3 \right\}.$$

(3) Consider the matrix equation $Ax = b$, where A is the matrix given by

$$A = \begin{bmatrix} 1 & 1 & 1 \\ 0 & 0 & 0 \\ 0 & 0 & 0 \end{bmatrix}$$

and $b \in \mathbb{F}^4$. This corresponds to a square inhomogeneous system of linear equations with two free variables. As above, this system has no solutions unless $b_2 = b_3 = 0$, and, given any scalars $\alpha, \beta \in \mathbb{F}$, every vector of the form

$$x = \begin{bmatrix} x_1 \\ x_2 \\ x_3 \end{bmatrix} = \begin{bmatrix} b_1 - \alpha - \beta \\ \alpha \\ \beta \end{bmatrix} = \begin{bmatrix} b_1 \\ 0 \\ 0 \end{bmatrix} + \alpha \begin{bmatrix} -1 \\ 1 \\ 0 \end{bmatrix} + \beta \begin{bmatrix} -1 \\ 0 \\ 1 \end{bmatrix} = u + \alpha n^{(1)} + \beta n^{(2)}$$

is a solution to $Ax = b$. Recall from Example A.3.10, that the solution space for the associated homogeneous matrix equation $Ax = 0$ is

$$N = \left\{ (x_1, x_2, x_3) \in \mathbb{F}^3 \mid x_1 + x_2 + x_3 = 0 \right\} = \text{span}\left((-1, 1, 0), (-1, 0, 1) \right).$$

Thus, in the language of Theorem A.3.12, we have that u is a particular solution for $Ax = b$ and that $(n^{(1)}, n^{(2)})$ is a basis for N. Therefore, the solution set for $Ax = b$ is

$$U = (b_1, 0, 0) + N = \left\{ (x_1, x_2, x_3) \in \mathbb{F}^3 \mid x_1 + x_2 + x_3 = b_1 \right\}.$$

A.3.4 *Solving linear systems with LU-factorization*

Let $n \in \mathbb{Z}_+$ be a positive integer, and suppose that $A \in \mathbb{F}^{n \times n}$ is an upper triangular matrix and that $b \in \mathbb{F}^n$ is a column vector. Then, in order to solve the matrix equation $Ax = b$, there is no need to apply Gaussian elimination. Instead, we can exploit the triangularity of A in order to directly obtain a solution.

Using the notation in System (A.1), note that the last equation in the linear system corresponding to $Ax = b$ can only involve the single unknown x_n, and so we can obtain the solution

$$x_n = \frac{b_n}{a_{nn}}$$

as long as $a_{nn} \neq 0$. If $a_{nn} = 0$, then we must be careful to distinguish between the two cases in which $b_n = 0$ or $b_n \neq 0$. Thus, for reasons that will become clear below, we assume that the diagonal elements of A are all non-zero. Under this assumption, there is no ambiguity in substituting the solution for x_n into the penultimate (a.k.a. second-to-last) equation. Since A is upper triangular, the penultimate equation involves only the single unknown x_{n-1}, and so we obtain the solution

$$x_{n-1} = \frac{b_{n-1} - a_{n-1,n} x_n}{a_{n-1,n-1}}.$$

We can then similarly substitute the solutions for x_n and x_{n-1} into the antepenultimate (a.k.a. third-to-last) equation in order to solve for x_{n-2}, and so on until a complete solution is found. In particular,

$$x_1 = \frac{b_1 - \sum_{k=2}^{n} a_{nk} x_k}{a_{11}}.$$

As in Example A.3.5, we call this process **back substitution**. Given an arbitrary linear system, back substitution essentially allows us to halt the Gaussian elimination procedure and immediately obtain a solution for the system as soon as an upper triangular matrix (possibly in REF or even RREF) has been obtained from the coefficient matrix.

A similar procedure can be applied when A is lower triangular. Again using the notation in System (A.1), the first equation contains only x_1, and so

$$x_1 = \frac{b_1}{a_{11}}.$$

We are again assuming that the diagonal entries of A are all non-zero. Then, acting similarly to back substitution, we can substitute the solution for x_1 into the second equation in order to obtain

$$x_2 = \frac{b_2 - a_{21}x_1}{a_{22}}.$$

Continuing this process, we have created a **forward substitution** procedure. In particular,

$$x_n = \frac{b_n - \sum_{k=1}^{n-1} a_{nk}x_k}{a_{nn}}.$$

More generally, suppose that $A \in \mathbb{F}^{n \times n}$ is an arbitrary square matrix for which there exists a lower triangular matrix $L \in \mathbb{F}^{n \times n}$ and an upper triangular matrix $U \in \mathbb{F}^{n \times n}$ such that $A = LU$. When such matrices exist, we call $A = LU$ an **LU-factorization** (a.k.a. **LU-decomposition**) of A. The benefit of such a factorization is that it allows us to exploit the triangularity of L and U when solving linear systems having coefficient matrix A.

To see this, suppose that $A = LU$ is an LU-factorization for the matrix $A \in \mathbb{F}^{n \times n}$ and that $b \in \mathbb{F}^n$ is a column vector. (As above, we also assume that none of the diagonal entries in either L or U is zero.) Furthermore, set $y = Ux$, where x is the as yet unknown solution of $Ax = b$. Then, by substitution, y must satisfy

$$Ly = b.$$

Then, since L is lower triangular, we can immediately solve for y via forward substitution. In other words, we are using the associativity of matrix multiplication (cf. Theorem A.2.6) in order to conclude that

$$(A)x = (LU)x = L(Ux) = L(y).$$

Then, once we have obtained $y \in \mathbb{F}^n$, we can apply back substitution in order to solve for x in the matrix equation

$$Ux = y.$$

In general, one can only obtain an LU-factorization for a matrix $A \in \mathbb{F}^{n \times n}$ when there exist elementary "row combination" matrices $E_1, E_2, \ldots, E_k \in \mathbb{F}^{n \times n}$ and an upper triangular matrix U such that

$$E_k E_{k-1} \cdots E_1 A = U.$$

There are various generalizations of LU-factorization that allow for more than just elementary "row combinations" matrices in this product, but we do not mention them here. Instead, we provide a detailed example that illustrates how to obtain an LU-factorization and then how to use such a factorization in solving linear systems.

Example A.3.15. Consider the matrix $A \in \mathbb{F}^{3\times3}$ given by

$$A = \begin{bmatrix} 2 & 3 & 4 \\ 4 & 5 & 10 \\ 4 & 8 & 2 \end{bmatrix}.$$

Using the techniques illustrated in Example A.3.5, we have the following matrix product:

$$\begin{bmatrix} 1 & 0 & 0 \\ 0 & 1 & 0 \\ 0 & 2 & 1 \end{bmatrix} \begin{bmatrix} 1 & 0 & 0 \\ 0 & 1 & 0 \\ -2 & 0 & 1 \end{bmatrix} \begin{bmatrix} 1 & 0 & 0 \\ -2 & 1 & 0 \\ 0 & 0 & 1 \end{bmatrix} \begin{bmatrix} 2 & 3 & 4 \\ 4 & 5 & 10 \\ 4 & 8 & 2 \end{bmatrix} = \begin{bmatrix} 2 & 3 & 4 \\ 0 & -1 & 3 \\ 0 & 0 & -2 \end{bmatrix} = U.$$

In particular, we have found three elementary "row combination" matrices, which, when multiplied by A, produce an upper triangular matrix U.

Now, in order to produce a lower triangular matrix L such that $A = LU$, we rely on two facts about lower triangular matrices. First of all, any lower triangular matrix with entirely non-zero diagonal is invertible, and, second, the product of lower triangular matrices is always lower triangular. (Cf. Theorem A.2.7.) More specifically, we have that

$$\begin{bmatrix} 2 & 3 & 4 \\ 4 & 5 & 10 \\ 4 & 8 & 2 \end{bmatrix} = \begin{bmatrix} 1 & 0 & 0 \\ -2 & 1 & 0 \\ 0 & 0 & 1 \end{bmatrix}^{-1} \begin{bmatrix} 1 & 0 & 0 \\ 0 & 1 & 0 \\ -2 & 0 & 1 \end{bmatrix}^{-1} \begin{bmatrix} 1 & 0 & 0 \\ 0 & 1 & 0 \\ 0 & 2 & 1 \end{bmatrix}^{-1} \begin{bmatrix} 2 & 3 & 4 \\ 0 & -1 & 3 \\ 0 & 0 & -2 \end{bmatrix},$$

where

$$\begin{bmatrix} 1 & 0 & 0 \\ -2 & 1 & 0 \\ 0 & 0 & 1 \end{bmatrix}^{-1} \begin{bmatrix} 1 & 0 & 0 \\ 0 & 1 & 0 \\ -2 & 0 & 1 \end{bmatrix}^{-1} \begin{bmatrix} 1 & 0 & 0 \\ 0 & 1 & 0 \\ 0 & 2 & 1 \end{bmatrix}^{-1} = \begin{bmatrix} 1 & 0 & 0 \\ 2 & 1 & 0 \\ 0 & 0 & 1 \end{bmatrix} \begin{bmatrix} 1 & 0 & 0 \\ 0 & 1 & 0 \\ 2 & 0 & 1 \end{bmatrix} \begin{bmatrix} 1 & 0 & 0 \\ 0 & 1 & 0 \\ 0 & -2 & 1 \end{bmatrix}$$

$$= \begin{bmatrix} 1 & 0 & 0 \\ 2 & 1 & 0 \\ 2 & -2 & 1 \end{bmatrix}.$$

We call the resulting lower triangular matrix L and note that $A = LU$, as desired.

Now, define x, y, and b by

$$x = \begin{bmatrix} x_1 \\ x_2 \\ x_3 \end{bmatrix}, y = \begin{bmatrix} y_1 \\ y_2 \\ y_3 \end{bmatrix}, \text{ and } b = \begin{bmatrix} 6 \\ 16 \\ 2 \end{bmatrix}.$$

Applying forward substitution to $Ly = b$, we obtain the solution

$$\left. \begin{aligned} y_1 &= b_1 & = 6 \\ y_2 &= b_2 - 2y_1 & = 4 \\ y_3 &= b_3 - 2y_1 + 2y_2 & = -2 \end{aligned} \right\}.$$

Then, given this unique solution y to $Ly = b$, we can apply backward substitution to $Ux = y$ in order to obtain

$$\left. \begin{aligned} 2x_1 &= y_1 - 3x_2 - 4x_3 & = 8 \\ -1x_2 &= y_2 & - 2x_3 & = 2 \\ -2x_3 &= y_3 & = -2 \end{aligned} \right\}.$$

It follows that the unique solution to $Ax = b$ is

$$\left.\begin{array}{r} x_1 = 4 \\ x_2 = -2 \\ x_3 = 1 \end{array}\right\}.$$

In summary, we have given an algorithm for solving any matrix equation $Ax = b$ in which $A = LU$, where L is lower triangular, U is upper triangular, and both L and U have nothing but non-zero entries along their diagonals.

We note in closing that the simple procedures of back and forward substitution can also be used for computing the inverses of lower and upper triangular matrices. E.g., the inverse $U = (u_{ij})$ of the matrix

$$U^{-1} = \begin{bmatrix} 2 & 3 & 4 \\ 0 & -1 & 3 \\ 0 & 0 & -2 \end{bmatrix}$$

must satisfy

$$\begin{bmatrix} 2u_{11} + 3u_{21} + 4u_{31} & 2u_{12} + 3u_{22} + 4u_{32} & 2u_{13} + 3u_{23} + 4u_{33} \\ -u_{21} + 3u_{31} & -u_{22} + 3u_{32} & -u_{23} + 3u_{33} \\ -2u_{31} & -2u_{32} & -2u_{33} \end{bmatrix} = U^{-1}U = I_3 = \begin{bmatrix} 1 & 0 & 0 \\ 0 & 1 & 0 \\ 0 & 0 & 1 \end{bmatrix},$$

from which we obtain the linear system

$$\left.\begin{array}{rcl} 2u_{11} +3u_{21} +4u_{31} &=& 1 \\ 2u_{12} +3u_{22} +4u_{32} &=& 0 \\ 2u_{13} +3u_{23} +4u_{33} &=& 0 \\ -u_{21} +3u_{31} &=& 0 \\ -u_{22} +3u_{32} &=& 1 \\ -u_{23} +3u_{33} &=& 0 \\ -2u_{31} &=& 0 \\ -2u_{32} &=& 0 \\ -2u_{33} &=& 1 \end{array}\right\}$$

in the nine variables $u_{11}, u_{12}, \ldots, u_{33}$. Since this linear system has an upper triangular coefficient matrix, we can apply back substitution in order to directly solve for the entries in U.

The only condition we imposed upon the triangular matrices above was that all diagonal entries were non-zero. Since the determinant of a triangular matrix is given by the product of its diagonal entries, this condition is necessary and sufficient for a triangular matrix to be non-singular. Moreover, once the inverses of both L and U in an LU-factorization have been obtained, we can immediately calculate the inverse for $A = LU$ by applying Theorem A.2.9(4):

$$A^{-1} = (LU)^{-1} = U^{-1}L^{-1}.$$

A.4 Matrices and linear maps

This section is devoted to illustrating how linear maps are a fundamental tool for gaining insight into the solutions to systems of linear equations with n unknowns. Using the tools of Linear Algebra, many familiar facts about systems with two unknowns can be generalized to an arbitrary number of unknowns without much effort.

A.4.1 *The canonical matrix of a linear map*

Let $m, n \in \mathbb{Z}_+$ be positive integers. Then, given a choice of bases for the vector spaces $\mathbb{F}^n$ and $\mathbb{F}^m$, there is a correspondence between matrices and linear maps. In other words, as discussed in Section 6.6, every linear map in the set $\mathcal{L}(\mathbb{F}^n, \mathbb{F}^m)$ uniquely corresponds to exactly one $m \times n$ matrix in $\mathbb{F}^{m \times n}$. However, you should not take this to mean that matrices and linear maps are interchangeable or indistinguishable ideas. By itself, a matrix in the set $\mathbb{F}^{m \times n}$ is nothing more than a collection of mn scalars that have been arranged in a rectangular shape. It is only when a matrix appears in a specific context in which a basis for the underlying vector space has been chosen, that the theory of linear maps becomes applicable. In particular, one can gain insight into the solutions of matrix equations when the coefficient matrix is viewed as the matrix associated to a linear map under a convenient choice of bases for $\mathbb{F}^n$ and $\mathbb{F}^m$.

Given a positive integer, $k \in \mathbb{Z}_+$, one particularly convenient choice of basis for $\mathbb{F}^k$ is the so-called **standard basis** (a.k.a. the **canonical basis**) $e_1, e_2, \ldots, e_k$, where each e_i is the k-tuple having zeros for each of its components other than in the ith position:

$$e_i = (0, 0, \ldots, 0, 1, 0, \ldots, 0).$$
$$\uparrow$$
$$i$$

Then, taking the vector spaces $\mathbb{F}^n$ and $\mathbb{F}^m$ with their canonical bases, we say that the matrix $A \in \mathbb{F}^{m \times n}$ associated to the linear map $T \in \mathcal{L}(\mathbb{F}^n, \mathbb{F}^m)$ is the **canonical matrix** for T. With this choice of bases we have

$$T(x) = Ax, \ \forall\, x \in \mathbb{F}^n. \tag{A.5}$$

In other words, one can compute the action of the linear map upon any vector in $\mathbb{F}^n$ by simply multiplying the vector by the associated canonical matrix A. There are many circumstances in which one might wish to use non-canonical bases for either $\mathbb{F}^n$ or $\mathbb{F}^m$, but the trade-off is that Equation (A.5) will no longer hold as stated. (To modify Equation (A.5) for use with non-standard bases, one needs to use coordinate vectors as described in Chapter 10.)

The utility of Equation (A.5) cannot be over-emphasized. To get a sense of this, consider once again the generic matrix equation (Equation (A.3))

$$Ax = b,$$

which involves a given matrix $A = (a_{ij}) \in \mathbb{F}^{m \times n}$, a given vector $b \in \mathbb{F}^m$, and the n-tuple of unknowns x. To provide a solution to this equation means to provide a vector $x \in \mathbb{F}^n$ for which the matrix product Ax is exactly the vector b. In light of Equation (A.5), the question of whether such a vector $x \in \mathbb{F}^n$ exists is equivalent to asking whether or not the vector b is in the range of the linear map T.

The encoding of System (A.1) into Equation (A.3) is more than a mere change of notation. The reinterpretation of Equation (A.3) using linear maps is a genuine change of viewpoint. Solving System (A.1) (and thus Equation (A.3)) essentially amounts to understanding how m distinct objects interact in an ambient space of dimension n. (In particular, solutions to System (A.1) correspond to the points of intersect of m hyperplanes in $\mathbb{F}^n$.) On the other hand, questions about a linear map involve understanding a single object, i.e., the linear map itself. Such a point of view is both extremely flexible and fruitful, as we illustrate in the next section.

A.4.2 *Using linear maps to solve linear systems*

Encoding a linear system as a matrix equation is more than just a notational trick. Perhaps most fundamentally, the resulting linear map viewpoint can then be used to provide clear insight into the exact structure of solutions to the original linear system. We illustrate this in the following series of revisited examples.

Example A.4.1. Consider the following inhomogeneous linear system from Example 1.2.1:

$$\left. \begin{array}{r} 2x_1 + x_2 = 0 \\ x_1 - x_2 = 1 \end{array} \right\},$$

where x_1 and x_2 are unknown real numbers. To solve this system, we can first form the matrix $A \in \mathbb{R}^{2 \times 2}$ and the column vector $b \in \mathbb{R}^2$ such that

$$A \begin{bmatrix} x_1 \\ x_2 \end{bmatrix} = \begin{bmatrix} 2 & 1 \\ 1 & -1 \end{bmatrix} \begin{bmatrix} x_1 \\ x_2 \end{bmatrix} = \begin{bmatrix} 0 \\ 1 \end{bmatrix} = b.$$

In other words, we have reinterpreted solving the original linear system as asking when the column vector

$$\begin{bmatrix} 2 & 1 \\ 1 & -1 \end{bmatrix} \begin{bmatrix} x_1 \\ x_2 \end{bmatrix} = \begin{bmatrix} 2x_1 + x_2 \\ x_1 - x_2 \end{bmatrix}$$

is equal to the column vector b. Equivalently, this corresponds to asking what input vector results in b being an element of the range of the linear map $T : \mathbb{R}^2 \to \mathbb{R}^2$ defined by

$$T \left(\begin{bmatrix} x_1 \\ x_2 \end{bmatrix} \right) = \begin{bmatrix} 2x_1 + x_2 \\ x_1 - x_2 \end{bmatrix}.$$

More precisely, T is the linear map having canonical matrix A.

It should be clear that b is in the range of T, since, from Example 1.2.1,

$$T \left(\begin{bmatrix} 1/3 \\ -2/3 \end{bmatrix} \right) = \begin{bmatrix} 0 \\ 1 \end{bmatrix}.$$

In addition, note that T is a bijective function. (This can be proven, for example, by noting that the canonical matrix A for T is invertible.) Since T is bijective, this means that

$$x = \begin{bmatrix} x_1 \\ x_2 \end{bmatrix} = \begin{bmatrix} 1/3 \\ -2/3 \end{bmatrix}$$

is the only possible input vector that can result in the output vector b, and so we have verified that x is the unique solution to the original linear system. Moreover, this technique can be trivially generalized to any number of equations.

Example A.4.2. Consider the matrix A and the column vectors x and b from Example A.3.5:

$$A = \begin{bmatrix} 2 & 5 & 3 \\ 1 & 2 & 3 \\ 1 & 0 & 8 \end{bmatrix}, \quad x = \begin{bmatrix} x_1 \\ x_2 \\ x_3 \end{bmatrix}, \quad \text{and} \quad b = \begin{bmatrix} 4 \\ 5 \\ 9 \end{bmatrix}.$$

Here, asking if the equation $Ax = b$ has a solution is equivalent to asking if b is an element of the range of the linear map $T : \mathbb{F}^3 \to \mathbb{F}^3$ defined by

$$T\left(\begin{bmatrix} x_1 \\ x_2 \\ x_3 \end{bmatrix} \right) = \begin{bmatrix} 2x_1 + 5x_2 + 3x_3 \\ x_1 + 2x_2 + 3x_3 \\ 2x_1 + 8x_3 \end{bmatrix}.$$

In order to answer this corresponding question regarding the range of T, we take a closer look at the following expression obtained in Example A.3.5:

$$A = E_0^{-1} E_1^{-1} \cdots E_7^{-1}.$$

Here, we have factored A into the product of eight elementary matrices. From the linear map point of view, this means that we can apply the results of Section 6.6 in order to obtain the factorization

$$T = S_0 \circ S_1 \circ \cdots \circ S_7,$$

where S_i is the (invertible) linear map having canonical matrix E_i^{-1} for $i = 0, \ldots, 7$.

This factorization of the linear map T into a composition of invertible linear maps furthermore implies that T itself is invertible. In particular, T is surjective, and so b must be an element of the range of T. Moreover, T is also injective, and so b has exactly one pre-image. Thus, the solution that was found for $Ax = b$ in Example A.3.5 is unique.

In the above examples, we used the bijectivity of a linear map in order to prove the uniqueness of solutions to linear systems. As discussed in Section A.3, many linear systems do not have unique solutions. Instead, there are exactly two other possibilities: if a linear system does not have a unique solution, then it will either have no solution or it will have infinitely many solutions. Fundamentally, this is because finding solutions to a linear system is equivalent to describing the pre-image (a.k.a. pullback) of an element in the codomain of a linear map.

In particular, based upon the discussion in Section A.3.2, it should be clear that solving a homogeneous linear system corresponds to describing the null space of some corresponding linear map. In other words, given any matrix $A \in \mathbb{F}^{m \times n}$, finding the solution space N to the matrix equation $Ax = 0$ (as defined in Section A.3.2) is the same thing as finding $\mathrm{null}(T)$, where $T \in \mathcal{L}(\mathbb{F}^n, \mathbb{F}^m)$ is the linear map having canonical matrix A. (Recall from Section 6.2 that $\mathrm{null}(T)$ is a subspace of $\mathbb{F}^n$.) Thus, the fact that every homogeneous linear system has the trivial solution then is equivalent to the fact that the image of the zero vector under any linear map always results in the zero vector, and determining whether or not the trivial solution is unique can be viewed as a dimensionality question about the null space of a corresponding linear map.

We close this section by illustrating this, along with the case for inhomogeneous systems, in the following examples.

Example A.4.3. The following examples use the same matrices as in Example A.3.10.

(1) Consider the matrix equation $Ax = b$, where A is the matrix given by

$$A = \begin{bmatrix} 1 & 0 & 0 \\ 0 & 1 & 0 \\ 0 & 0 & 1 \\ 0 & 0 & 0 \end{bmatrix}$$

and $b \in \mathbb{F}^4$ is a column vector. Here, asking if this matrix equation has a solution corresponds to asking if b is an element of the range of the linear map $T : \mathbb{F}^3 \to \mathbb{F}^4$ defined by

$$T\left(\begin{bmatrix} x_1 \\ x_2 \\ x_3 \end{bmatrix} \right) = \begin{bmatrix} x_1 \\ x_2 \\ x_3 \\ 0 \end{bmatrix}.$$

From the linear map point of view, it should be clear that $Ax = b$ has a solution if and only if the fourth component of b is zero. In particular, T is not surjective, so $Ax = b$ cannot have a solution for every possible choice of b.

However, it should also be clear that T is injective, from which $\mathrm{null}(T) = \{0\}$. Thus, when $b = 0$, the homogeneous matrix equation $Ax = 0$ has only the trivial solution, and so we can apply Theorem A.3.12 in order to verify that $Ax = b$ has a unique solution for any b having fourth component equal to zero.

(2) Consider the matrix equation $Ax = b$, where A is the matrix given by

$$A = \begin{bmatrix} 1 & 0 & 1 \\ 0 & 1 & 1 \\ 0 & 0 & 0 \\ 0 & 0 & 0 \end{bmatrix}$$

and $b \in \mathbb{F}^4$ is a column vector. Here, asking if this matrix equation has a solution corresponds to asking if b is an element of the range of the linear map $T : \mathbb{F}^3 \to \mathbb{F}^4$ defined by

$$T\left(\begin{bmatrix} x_1 \\ x_2 \\ x_3 \end{bmatrix}\right) = \begin{bmatrix} x_1 + x_3 \\ x_2 + x_3 \\ 0 \\ 0 \end{bmatrix}.$$

From the linear map point of view, it should be clear that $Ax = b$ has a solution if and only if the third and fourth components of b are zero. In particular, $2 = \dim(\text{range}\,(T)) < \dim(\mathbb{F}^4) = 4$ so that T cannot be surjective, and so $Ax = b$ cannot have a solution for every possible choice of b.

In addition, it should also be clear that T is not injective. E.g.,

$$T\left(\begin{bmatrix} -1 \\ -1 \\ 1 \end{bmatrix}\right) = \begin{bmatrix} 0 \\ 0 \\ 0 \\ 0 \end{bmatrix}.$$

Thus, $\{0\} \subsetneq \text{null}(T)$, and so the homogeneous matrix equation $Ax = 0$ will necessarily have infinitely many solutions since $\dim(\text{null}(T)) > 0$. Using the Dimension Formula,

$$\dim(\text{null}(T)) = \dim(\mathbb{F}^3) - \dim(\text{range}\,(T)) = 3 - 2 = 1,$$

and so the solution space for $Ax = 0$ is a one-dimensional subspace of $\mathbb{F}^3$. Moreover, by applying Theorem A.3.12, we see that $Ax = b$ must then also have infinitely many solutions for any b having third and fourth components equal to zero.

(3) Consider the matrix equation $Ax = b$, where A is the matrix given by

$$A = \begin{bmatrix} 1 & 1 & 1 \\ 0 & 0 & 0 \\ 0 & 0 & 0 \end{bmatrix}$$

and $b \in \mathbb{F}^3$ is a column vector. Here, asking if this matrix equation has a solution corresponds to asking if b is an element of the range of the linear map $T : \mathbb{F}^3 \to \mathbb{F}^3$ defined by

$$T\left(\begin{bmatrix} x_1 \\ x_2 \\ x_3 \end{bmatrix}\right) = \begin{bmatrix} x_1 + x_2 + x_3 \\ 0 \\ 0 \end{bmatrix}.$$

From the linear map point of view, it should be extremely clear that $Ax = b$ has a solution if and only if the second and third components of b are zero. In particular, $1 = \dim(\text{range}\,(T)) < \dim(\mathbb{F}^3) = 3$ so that T cannot be surjective, and so $Ax = b$ cannot have a solution for every possible choice of b.

In addition, it should also be clear that T is not injective. E.g.,

$$T\left(\begin{bmatrix} 1/2 \\ 1/2 \\ -1 \end{bmatrix}\right) = \begin{bmatrix} 0 \\ 0 \\ 0 \end{bmatrix}.$$

Thus, $\{0\} \subsetneq \text{null}(T)$, and so the homogeneous matrix equation $Ax = 0$ will necessarily have infinitely many solutions since $\dim(\text{null}(T)) > 0$. Using the Dimension Formula,

$$\dim(\text{null}(T)) = \dim(\mathbb{F}^3) - \dim(\text{range}\,(T)) = 3 - 1 = 2,$$

and so the solution space for $Ax = 0$ is a two-dimensional subspace of $\mathbb{F}^3$. Moreover, by applying Theorem A.3.12, we see that $Ax = b$ must then also have infinitely many solutions for any b having second and third components equal to zero.

A.5 Special operations on matrices

In this section, we define three important operations on matrices called the transpose, conjugate transpose, and the trace. These will then be seen to interact with matrix multiplication and invertibility in order to form special classes of matrices that are extremely important to applications of Linear Algebra.

A.5.1 *Transpose and conjugate transpose*

Given positive integers $m, n \in \mathbb{Z}_+$ and any matrix $A = (a_{ij}) \in \mathbb{F}^{m \times n}$, we define the **transpose** $A^T = ((a^T)_{ij}) \in \mathbb{F}^{n \times m}$ and the **conjugate transpose** $A^* = ((a^*)_{ij}) \in \mathbb{F}^{n \times m}$ by

$$(a^T)_{ij} = a_{ji} \text{ and } (a^*)_{ij} = \overline{a_{ji}},$$

where $\overline{a_{ji}}$ denotes the complex conjugate of the scalar $a_{ji} \in \mathbb{F}$. In particular, if $A \in \mathbb{R}^{m \times n}$, then note that $A^T = A^*$.

Example A.5.1. With notation as in Example A.1.3,

$$A^T = \begin{bmatrix} 3 & -1 & 1 \end{bmatrix}, B^T = \begin{bmatrix} 4 & 0 \\ -1 & 2 \end{bmatrix}, C^T = \begin{bmatrix} 1 \\ 4 \\ 2 \end{bmatrix},$$

$$D^T = \begin{bmatrix} 1 & -1 & 3 \\ 5 & 0 & 2 \\ 2 & 1 & 4 \end{bmatrix}, \text{ and } E^T = \begin{bmatrix} 6 & -1 & 4 \\ 1 & 1 & 1 \\ 3 & 2 & 3 \end{bmatrix}.$$

One of the motivations for defining the operations of transpose and conjugate transpose is that they interact with the usual arithmetic operations on matrices in a natural manner. We summarize the most fundamental of these interactions in the following theorem.

Theorem A.5.2. *Given positive integers $m, n \in \mathbb{Z}_+$ and any matrices $A, B \in \mathbb{F}^{m \times n}$,*

(1) $(A^T)^T = A$ and $(A^*)^* = A$.
(2) $(A + B)^T = A^T + B^T$ and $(A + B)^* = A^* + B^*$.
(3) $(\alpha A)^T = \alpha A^T$ and $(\alpha A)^* = \overline{\alpha} A^*$, where $\alpha \in \mathbb{F}$ is any scalar.
(4) $(AB)^T = B^T A^T$.
(5) if $m = n$ and $A \in GL(n, \mathbb{F})$, then $A^T, A^* \in GL(n, \mathbb{F})$ with respective inverses given by
$$(A^T)^{-1} = (A^{-1})^T \quad and \quad (A^*)^{-1} = (A^{-1})^*.$$

Another motivation for defining the transpose and conjugate transpose operations is that they allow us to define several very special classes of matrices.

Definition A.5.3. Given a positive integer $n \in \mathbb{Z}_+$, we say that the square matrix $A \in \mathbb{F}^{n \times n}$

(1) is **symmetric** if $A = A^T$.
(2) is **Hermitian** if $A = A^*$.
(3) is **orthogonal** if $A \in GL(n, \mathbb{R})$ and $A^{-1} = A^T$. Moreover, we define the **(real) orthogonal group** to be the set $O(n) = \{A \in GL(n, \mathbb{R}) \mid A^{-1} = A^T\}$.
(4) is **unitary** if $A \in GL(n, \mathbb{C})$ and $A^{-1} = A^*$. Moreover, we define the **(complex) unitary group** to be the set $U(n) = \{A \in GL(n, \mathbb{C}) \mid A^{-1} = A^*\}$.

A lot can be said about these classes of matrices. Both $O(n)$ and $U(n)$, for example, form a group under matrix multiplication. Additionally, real symmetric and complex Hermitian matrices always have real eigenvalues. Moreover, given any matrix $A \in \mathbb{R}^{m \times n}$, AA^T is a symmetric matrix with real, non-negative eigenvalues. Similarly, for $A \in \mathbb{C}^{m \times n}$, AA^* is Hermitian with real, non-negative eigenvalues.

A.5.2 *The trace of a square matrix*

Given a positive integer $n \in \mathbb{Z}_+$ and any square matrix $A = (a_{ij}) \in \mathbb{F}^{n \times n}$, we define the **trace** of A to be the scalar
$$\text{trace}(A) = \sum_{k=1}^{n} a_{kk} \in \mathbb{F}.$$

Example A.5.4. With notation as in Example A.1.3 above,

$\text{trace}(B) = 4 + 2 = 6$, $\text{trace}(D) = 1 + 0 + 4 = 5$, and $\text{trace}(E) = 6 + 1 + 3 = 10$.

Note, in particular, that the traces of A and C are not defined since these are not square matrices.

We summarize some of the most basic properties of the trace operation in the following theorem, including its connection to the transpose operations defined in the previous section.

Theorem A.5.5. *Given positive integers $m, n \in \mathbb{Z}_+$ and square matrices $A, B \in \mathbb{F}^{n \times n}$,*

(1) *$\operatorname{trace}(\alpha A) = \alpha \operatorname{trace}(A)$, for any scalar $\alpha \in \mathbb{F}$.*
(2) *$\operatorname{trace}(A + B) = \operatorname{trace}(A) + \operatorname{trace}(B)$.*
(3) *$\operatorname{trace}(A^T) = \operatorname{trace}(A)$ and $\operatorname{trace}(A^*) = \overline{\operatorname{trace}(A)}$.*
(4) *$\operatorname{trace}(AA^*) = \sum_{k=1}^{n} \sum_{\ell=1}^{n} |a_{k\ell}|^2$. In particular, $\operatorname{trace}(AA^*) = 0$ if and only if $A = 0_{n \times n}$.*
(5) *$\operatorname{trace}(AB) = \operatorname{trace}(BA)$. More generally, given matrices $A_1, \ldots, A_m \in \mathbb{F}^{n \times n}$, the trace operation has the so-called **cyclic property**, meaning that*

$$\operatorname{trace}(A_1 \cdots A_m) = \operatorname{trace}(A_2 \cdots A_m A_1) = \cdots = \operatorname{trace}(A_m A_1 \cdots A_{m-1}).$$

Moreover, if we define a linear map $T : \mathbb{F}^n \to \mathbb{F}^n$ by setting $T(v) = Av$ for each $v \in \mathbb{F}^n$ and if T has distinct eigenvalues $\lambda_1, \ldots, \lambda_n$, then $\operatorname{trace}(A) = \sum_{k=1}^{n} \lambda_k$.

Exercises for Appendix A

Calculational Exercises

(1) In each of the following, find matrices A, x, and b such that the given system of linear equations can be expressed as the single matrix equation $Ax = b$.

(a)
$$\left. \begin{array}{r} 2x_1 - 3x_2 + 5x_3 = 7 \\ 9x_1 - x_2 + x_3 = -1 \\ x_1 + 5x_2 + 4x_3 = 0 \end{array} \right\}$$

(b)
$$\left. \begin{array}{r} 4x_1 - 3x_3 + x_4 = 1 \\ 5x_1 + x_2 - 8x_4 = 3 \\ 2x_1 - 5x_2 + 9x_3 - x_4 = 0 \\ 3x_2 - x_3 + 7x_4 = 2 \end{array} \right\}$$

(2) In each of the following, express the matrix equation as a system of linear equations.

(a)
$$\begin{bmatrix} 3 & -1 & 2 \\ 4 & 3 & 7 \\ -2 & 1 & 5 \end{bmatrix} \begin{bmatrix} x_1 \\ x_2 \\ x_3 \end{bmatrix} = \begin{bmatrix} 2 \\ -1 \\ 4 \end{bmatrix}$$

(b)
$$\begin{bmatrix} 3 & -2 & 0 & 1 \\ 5 & 0 & 2 & -2 \\ 3 & 1 & 4 & 7 \\ -2 & 5 & 1 & 6 \end{bmatrix} \begin{bmatrix} w \\ x \\ y \\ z \end{bmatrix} = \begin{bmatrix} 0 \\ 0 \\ 0 \\ 0 \end{bmatrix}$$

(3) Suppose that A, B, C, D, and E are matrices over $\mathbb{F}$ having the following sizes:

$$A \text{ is } 4 \times 5, \quad B \text{ is } 4 \times 5, \quad C \text{ is } 5 \times 2, \quad D \text{ is } 4 \times 2, \quad E \text{ is } 5 \times 4.$$

Determine whether the following matrix expressions are defined, and, for those that are defined, determine the size of the resulting matrix.

(a) BA (b) $AC + D$ (c) $AE + B$ (d) $AB + B$ (e) $E(A + B)$ (f) $E(AC)$

(4) Suppose that A, B, C, D, and E are the following matrices:

$$A = \begin{bmatrix} 3 & 0 \\ -1 & 2 \\ 1 & 1 \end{bmatrix}, \; B = \begin{bmatrix} 4 & -1 \\ 0 & 2 \end{bmatrix}, \; C = \begin{bmatrix} 1 & 4 & 2 \\ 3 & 1 & 5 \end{bmatrix},$$

$$D = \begin{bmatrix} 1 & 5 & 2 \\ -1 & 0 & 1 \\ 3 & 2 & 4 \end{bmatrix}, \; \text{and} \; E = \begin{bmatrix} 6 & 1 & 3 \\ -1 & 1 & 2 \\ 4 & 1 & 3 \end{bmatrix}.$$

Determine whether the following matrix expressions are defined, and, for those that are defined, compute the resulting matrix.

(a) $D + E$ (b) $D - E$ (c) $5A$ (d) $-7C$ (e) $2B - C$
(f) $2E - 2D$ (g) $-3(D + 2E)$ (h) $A - A$ (i) AB (j) BA
(k) $(3E)D$ (l) $(AB)C$ (m) $A(BC)$ (n) $(4B)C + 2B$ (o) $D - 3E$
(p) $CA + 2E$ (q) $4E - D$ (r) DD

(5) Suppose that A, B, and C are the following matrices and that $a = 4$ and $b = -7$.

$$A = \begin{bmatrix} 1 & 5 & 2 \\ -1 & 0 & 1 \\ 3 & 2 & 4 \end{bmatrix}, \; B = \begin{bmatrix} 6 & 1 & 3 \\ -1 & 1 & 2 \\ 4 & 1 & 3 \end{bmatrix}, \; \text{and} \; C = \begin{bmatrix} 1 & 5 & 2 \\ -1 & 0 & 1 \\ 3 & 2 & 4 \end{bmatrix}.$$

Verify computationally that

(a) $A + (B + C) = (A + B) + C$ (b) $(AB)C = A(BC)$
(c) $(a + b)C = aC + bC$ (d) $a(B - C) = aB - aC$
(e) $a(BC) = (aB)C = B(aC)$ (f) $A(B - C) = AB - AC$
(g) $(B + C)A = BA + CA$ (h) $a(bC) = (ab)C$
(i) $B - C = -C + B$

(6) Suppose that A is the matrix

$$A = \begin{bmatrix} 3 & 1 \\ 2 & 1 \end{bmatrix}.$$

Compute $p(A)$, where $p(z)$ is given by

(a) $p(z) = z - 2$ (b) $p(z) = 2z^2 - z + 1$
(c) $p(z) = z^3 - 2z + 4$ (d) $p(z) = z^2 - 4z + 1$

(7) Define matrices A, B, C, D, and E by

$$A = \begin{bmatrix} 3 & 1 \\ 2 & 1 \end{bmatrix}, \; B = \begin{bmatrix} 4 & -1 \\ 0 & 2 \end{bmatrix}, \; C = \begin{bmatrix} 2 & -3 & 5 \\ 9 & -1 & 1 \\ 1 & 5 & 4 \end{bmatrix},$$

$$D = \begin{bmatrix} 1 & 5 & 2 \\ -1 & 0 & 1 \\ 3 & 2 & 4 \end{bmatrix}, \; \text{and} \; E = \begin{bmatrix} 6 & 1 & 3 \\ -1 & 1 & 2 \\ 4 & 1 & 3 \end{bmatrix}.$$

(a) Factor each matrix into a product of elementary matrices and an RREF matrix.

(b) Find, if possible, the LU-factorization of each matrix.

(c) Determine whether or not each of these matrices is invertible, and, if possible, compute the inverse.

(8) Suppose that A, B, C, D, and E are the following matrices:

$$A = \begin{bmatrix} 3 & 0 \\ -1 & 2 \\ 1 & 1 \end{bmatrix}, \ B = \begin{bmatrix} 4 & -1 \\ 0 & 2 \end{bmatrix}, \ C = \begin{bmatrix} 1 & 4 & 2 \\ 3 & 1 & 5 \end{bmatrix},$$

$$D = \begin{bmatrix} 1 & 5 & 2 \\ -1 & 0 & 1 \\ 3 & 2 & 4 \end{bmatrix}, \ \text{and} \ E = \begin{bmatrix} 6 & 1 & 3 \\ -1 & 1 & 2 \\ 4 & 1 & 3 \end{bmatrix}.$$

Determine whether the following matrix expressions are defined, and, for those that are defined, compute the resulting matrix.

(a) $2A^T + C$ (b) $D^T - E^T$ (c) $(D - E)^T$

(d) $B^T + 5C^T$ (e) $\frac{1}{2}C^T - \frac{1}{4}A$ (f) $B - B^T$

(g) $3E^T - 3D^T$ (h) $(2E^T - 3D^T)^T$ (i) CC^T

(j) $(DA)^T$ (k) $(C^T B)A^T$ (l) $(2D^T - E)A$

(m) $(BA^T - 2C)^T$ (n) $B^T(CC^T - A^T A)$ (o) $D^T E^T - (ED)^T$

(p) $\text{trace}(DD^T)$ (q) $\text{trace}(4E^T - D)$ (r) $\text{trace}(C^T A^T + 2E^T)$

Proof-Writing Exercises

(1) Let $n \in \mathbb{Z}_+$ be a positive integer and $a_{i,j} \in \mathbb{F}$ be scalars for $i, j = 1, \ldots, n$. Prove that the following two statements are equivalent:

(a) The trivial solution $x_1 = \cdots = x_n = 0$ is the only solution to the homogeneous system of equations

$$\left. \begin{aligned} \sum_{k=1}^{n} a_{1,k} x_k &= 0 \\ &\vdots \\ \sum_{k=1}^{n} a_{n,k} x_k &= 0 \end{aligned} \right\}.$$

(b) For every choice of scalars $c_1, \ldots, c_n \in \mathbb{F}$, there is a solution to the system of equations

$$\left. \begin{aligned} \sum_{k=1}^{n} a_{1,k} x_k &= c_1 \\ &\vdots \\ \sum_{k=1}^{n} a_{n,k} x_k &= c_n \end{aligned} \right\}.$$

(2) Let A and B be any matrices.

  (a) Prove that if both AB and BA are defined, then AB and BA are both square matrices.

  (b) Prove that if A has size $m \times n$ and ABA is defined, then B has size $n \times m$.

(3) Suppose that A is a matrix satisfying $A^T A = A$. Prove that A is then a symmetric matrix and that $A = A^2$.

(4) Suppose A is an upper triangular matrix and that $p(z)$ is any polynomial. Prove or give a counterexample: $p(A)$ is a upper triangular matrix.

Appendix B

The Language of Sets and Functions

All of mathematics can be seen as the study of relations between collections of objects by rigorous rational arguments. More often than not the patterns in those collections and their relations are more important than the nature of the objects themselves. The power of mathematics has a lot to do with bringing patterns to the forefront and abstracting from the "real" nature of the objects. In mathematics, the collections are usually called *sets* and the objects are called the *elements* of the set. *Functions* are the most common type of relation between sets and their elements. It is therefore important to develop a good understanding of sets and functions and to know the vocabulary used to define sets and functions and to discuss their properties.

B.1   Sets

A **set** is an unordered collection of distinct objects, which we call its **elements**. A set is uniquely determined by its elements. If an object a is an element of a set A, we write $a \in A$, and say that a belongs to A or that A contains a. The negation of this statement is written as $a \notin A$, i.e., a is *not* an element of A. Note that both statements cannot be true at the same time.

If A and B are sets, they are identical (this means one and the same set), which we write as $A = B$, if they have exactly the same elements. In other words, $A = B$ if and only if for all $a \in A$ we have $a \in B$, and for all $b \in B$ we have $b \in A$. Equivalently, $A \neq B$ if and only if there is a difference in their elements: there exists $a \in A$ such that $a \notin B$ or there exists $b \in B$ such that $b \notin A$.

Example B.1.1. We start with the simplest examples of sets.

(1) The **empty set** (a.k.a. the **null set**), is what it sounds like: the set with *no* elements. We usually denote it by $\emptyset$ or sometimes by $\{ \}$. The empty set, $\emptyset$, is uniquely determined by the property that for all x we have $x \notin \emptyset$. Clearly, there is exactly one empty set.

(2) Next up are the **singletons**. A singleton is a set with exactly one element. If

171

that element is x we often write the singleton containing x as $\{x\}$. In spoken language, 'the singleton x' actually means the set $\{x\}$ and should always be distinguished from the element x: $x \neq \{x\}$. A set can be an element of another set but no set is an element of itself (more precisely, we adopt this as an axiom). E.g., $\{\{x\}\}$ is the singleton of which the unique element is the singleton $\{x\}$. In particular we also have $\{x\} \neq \{\{x\}\}$.

(3) One standard way of denoting sets is by listing its elements. E.g., the set $\{\alpha, \beta, \gamma\}$ contains the first three lower case Greek letters. The set is completely determined by what is in the list. The order in which the elements are listed is irrelevant. So, we have $\{\alpha, \gamma, \beta\} = \{\gamma, \beta, \alpha\} = \{\alpha, \beta, \gamma\}$, etc. Since a set cannot contain the same element twice (elements are distinct) the only reasonable meaning of something like $\{\alpha, \beta, \alpha, \gamma\}$ is that it is the same as $\{\alpha, \beta, \gamma\}$. Since $x \neq \{x\}$, $\{x, \{x\}\}$ is a set with two elements. Anything can be considered as an element of a set and there is not any kind of relation required between the elements in a set. E.g., the word 'apple' and the element uranium and the planet Pluto can be the three elements of a set. There is no restriction on the number of different sets a given element can belong to, except for the rule that a set cannot be an element of itself.

(4) The number of elements in a set may be infinite. E.g., $\mathbb{Z}$, $\mathbb{R}$, and $\mathbb{C}$, denote the sets of all integer, real, and complex numbers, respectively. It is not required that we can list all elements.

When introducing a new set (new for the purpose of the discussion at hand) it is crucial to define it unambiguously. It is not required that from a given definition of a set A, it is easy to determine what the elements of A are, or even how many there are, but it should be clear that, in principle, there is a unique and unambiguous answer to each question of the form "is x an element of A?". There are several common ways to define sets. Here are a few examples.

Example B.1.2.

(1) The simplest way is a generalization of the list notation to infinite lists that can be described by a pattern. E.g., the set of positive integers $\mathbb{N} = \{1, 2, 3, \ldots\}$. The list can be allowed to be bi-directional, as in the set of all integers $\mathbb{Z} = \{\ldots, -2, -1, 0, 1, 2, \ldots\}$. Note the use of triple dots $\ldots$ to indicate the continuation of the list.

(2) The so-called **set builder notation** gives more options to describe the membership of a set. E.g., the set of all even integers, often denote by $2\mathbb{Z}$, is defined by

$$2\mathbb{Z} = \{2a \mid a \in \mathbb{Z}\}.$$

Instead of the vertical bar, $|$, a colon, $:$, is also commonly used. For example, the open interval of the real numbers strictly between 0 and 1 is defined by

$$(0,1) = \{x \in \mathbb{R} : 0 < x < 1\}.$$

B.2   Subset, union, intersection, and Cartesian product

Definition B.2.1. Let A and B be sets. B is a **subset** of A, denoted by $B \subset A$, if and only if for all $b \in B$ we have $b \in A$. If $B \subset A$ and $B \neq A$, we say that B is a **proper subset** of A.

If $B \subset A$, one also says that B is **contained in** A, or that A contains B, which is sometimes denoted by $A \supset B$. The relation $\subset$ is called **inclusion**. If B is a proper subset of A the inclusion is said to be **strict**. To emphasize that an inclusion is not necessarily strict, the notation $B \subseteq A$ can be used but note that its mathematical meaning is identical to $B \subset A$. Strict inclusion is sometimes denoted by $B \subsetneq A$, but this is less common.

Example B.2.2. The following relations between sets are easy to verify:

(1) We have $\mathbb{N} \subset \mathbb{Z} \subset \mathbb{Q} \subset \mathbb{R} \subset \mathbb{C}$, and all these inclusions are strict.
(2) For any set A, we have $\emptyset \subset A$, and $A \subset A$.
(3) $(0, 1] \subset (0, 2)$.
(4) For $0 < a \leq b$, $[-a, a] \subset [-b, b]$. The inclusion is strict if $a < b$.

In addition to constructing sets directly, sets can also be obtained from other sets by a number of standard operations. The following definition introduces the basic operations of taking the **union**, **intersection**, and **difference** of sets.

Definition B.2.3. Let A and B be sets. Then

(1) The **union** of A and B, denoted by $A \cup B$, is defined by

$$A \cup B = \{x \mid x \in A \text{ or } x \in B\}.$$

(2) The **intersection** of A and B, denoted by $A \cap B$, is defined by

$$A \cap B = \{x \mid x \in A \text{ and } x \in B\}.$$

(3) The **set difference** of B from A, denoted by $A \setminus B$, is defined by

$$A \setminus B = \{x \mid x \in A \text{ and } x \notin B\}.$$

Often, the context provides a 'universe' of all possible elements pertinent to a given discussion. Suppose, we have given such a set of 'all' elements and let us call it U. Then, the **complement** of a set A, denoted by A^c, is defined as $A^c = U \setminus A$. In the following theorem the existence of a universe U is tacitly assumed.

Theorem B.2.4. *Let A, B, and C be sets. Then*

(1) (distributivity) $A \cap (B \cup C) = (A \cap B) \cup (A \cap C)$ *and* $A \cup (B \cap C) = (A \cup B) \cap (A \cup C)$.
(2) (De Morgan's Laws) $(A \cup B)^c = A^c \cap B^c$ *and* $(A \cap B)^c = A^c \cup B^c$.
(3) (relative complements) $A \setminus (B \cup C) = (A \setminus B) \cap (A \setminus C)$ *and* $A \setminus (B \cap C) = (A \setminus B) \cup (A \setminus C)$.

To familiarize yourself with the basic properties of sets and the basic operations of sets, it is a good exercise to write proofs for the three properties stated in the theorem.

The so-called **Cartesian product** of sets is a powerful and ubiquitous method to construct new sets out of old ones.

Definition B.2.5. Let A and B be sets. Then the **Cartesian product** of A and B, denoted by $A \times B$, is the set of all ordered pairs (a, b), with $a \in A$ and $b \in B$. In other words,

$$A \times B = \{(a, b) \mid a \in A, b \in B\}.$$

An important example of this construction is the Euclidean plane $\mathbb{R}^2 = \mathbb{R} \times \mathbb{R}$. It is not an accident that x and y in the pair (x, y) are called the *Cartesian* coordinates of the point (x, y) in the plane.

B.3   Relations

In this section we introduce two important types of relations: order relations and equivalence relations. A **relation** R between elements of a set A and elements of a set B is a subset of their Cartesian product: $R \subset A \times B$. When $A = B$, we also call R simply a relation on A.

Let A be a set and R a relation on A. Then,

- R is called **reflexive** if for all $a \in A$, $(a, a) \in R$.
- R is called **symmetric** if for all $a, b \in A$, if $(a, b) \in R$ then $(b, a) \in R$.
- R is called **antisymmetric** if for all $a, b \in A$ such that $(a, b) \in R$ and $(b, a) \in R$, $a = b$.
- R is called **transitive** if for all $a, b, c \in A$ such $(a, b) \in R$ and $(b, c) \in R$, we have $(a, c) \in R$.

Definition B.3.1. Let R be a relation on a set A. R is an **order relation** if R is *reflexive, antisymmetric, and transitive*. R is an **equivalence relation** if R is *reflexive, symmetric, and transitive.*

The notion of subset is an example of an order relation. To see this, first define the **power set** of a set A as the set of all its subsets. It is often denoted by $\mathcal{P}(A)$. So, for any set A, $\mathcal{P}(A) = \{B : B \subset A\}$. Then, the inclusion relation is defined as the relation R by setting

$$R = \{(B, C) \in \mathcal{P}(A) \times \mathcal{P}(A) \mid B \subset C\}.$$

Important relations, such as the subset relation, are given a convenient notation of the form $a <symbol> b$, to denote $(a, b) \in R$. The symbol for the inclusion relation is $\subset$.

Proposition B.3.2. *Inclusion is an order relation. Explicitly,*

(1) (reflexive) For all $B \in \mathcal{P}(A)$, $B \subset B$.
(2) (antisymmetric) For all $B, C \in \mathcal{P}(A)$, if $B \subset C$ and $C \subset B$, then $B = C$.
(3) (transitive) For all $B, C, D \in \mathcal{P}(A)$, if $B \subset C$ and $C \subset D$, then $B \subset D$.

Write a proof of this proposition as an exercise.

For any relation $R \subset A \times B$, the **inverse relation**, $R^{-1} \subset B \times A$, is defined by

$$R^{-1} = \{(b, a) \in B \times A \mid (a, b) \in R\}.$$

B.4   Functions

Let A and B be sets. A **function** with **domain** A and **codomain** B, denoted by
$f : A \to B$, is a relation between the elements of A and B satisfying the properties:
for all $a \in A$, there is a *unique* $b \in B$ such that $(a, b) \in f$. The symbol used to
denote a function as a relation is an arrow: $(a, b) \in f$ is written as $a \to b$ (often
also $a \mapsto b$). It is not necessary, and a bit cumbersome, to remind ourselves that
functions are a special kind of relation and a more convenient notation is used all
the time: $f(a) = b$. If f is a function we then have, by definition, $f(a) = b$ and
$f(a) = c$ implies $b = c$. In other words, for each $a \in A$, there is exactly one $b \in B$
such that $f(a) = b$. b is called the **image** of a under f. When A and B are sets of
numbers, a is sometimes referred to as the **argument** of the function and $b = f(a)$
is often referred to as the **value** of f in a.

The requirement that there is an image $b \in B$ for *all* $a \in A$ is sometimes relaxed
in the sense that the domain of the function is a, sometimes not explicitly specified,
subset of A. It is important to remember, however, that a function is not properly
defined unless we have also given its domain.

When we consider the **graph** of a function, we are relying on the definition of a
function as a relation. The graph G of a function $f : A \to B$ is the subset of $A \times B$
defined by

$$G = \{(a, f(a)) \mid a \in A\}.$$

The **range** of a function $f : A \to B$, denoted by range (f), or also $f(A)$, is the
subset of its codomain consisting of all $b \in B$ that are the image of some $a \in A$:

$$\text{range}\,(f) = \{b \in B \mid \text{ there exists } a \in A \text{ such that } f(a) = b\}.$$

The **pre-image** of $b \in B$ is the subset of all $a \in A$ that have b as their image. This
subset is often denoted by $f^{-1}(b)$.

$$f^{-1}(b) = \{a \in A \mid f(a) = b\}.$$

Note that $f^{-1}(b) = \emptyset$ if and only if $b \in B \setminus \text{range}\,(f)$.

Functions of various kinds are ubiquitous in mathematics and a large vocabulary
has developed, some of which is redundant. The term *map* is often used as an
alternative for function and when the domain and codomain coincide the term

transformation is often used instead of function. There is a large number of terms for functions in a particular context with special properties. The three most basic properties are given in the following definition.

Definition B.4.1. Let $f : A \to B$ be a function. Then we call f

(1) **injective** (f is an **injection**) if $f(a) = f(b)$ implies $a = b$. In other words, no two elements of the domain have the same image. An injective function is also called **one-to-one**.

(2) **surjective** (f is a **surjection** if range $(f) = B$. In other words, each $b \in B$ is the image of at least one $a \in A$. Such a function is also called **onto**.

(3) **bijective** (f is a **bijection**) if f is both injective and surjective, i.e., **one-to-one and onto**. This means that f gives a one-to-one correspondence between *all* elements of A and *all* elements of B.

Let $f : A \to B$ and $g : B \to C$ be two functions so that the codomain of f coincides with the domain of g. Then, the **composition** 'g after f', denoted by $g \circ f$, is the function $g \circ f : A \to C$, defined by $a \mapsto g(f(a))$.

For every set A, we define the **identity map**, which we will denote here by id_A or id for short. $\mathrm{id}_A : A \to A$ is defined by $\mathrm{id}_A(a) = a$ for all $a \in A$. Clearly, id_A is a bijection.

If f is a bijection, it is invertible, i.e., the inverse relation is also a function, denoted by f^{-1}. It is the unique bijection $B \to A$ such that $f^{-1} \circ f = \mathrm{id}_A$ and $f \circ f^{-1} = \mathrm{id}_B$.

Proposition B.4.2. *Let $f : A \to B$ and $g : B \to C$ be bijections. Then, their composition $g \circ f$ is a bijection and*

$$(g \circ f)^{-1} = f^{-1} \circ g^{-1}.$$

Prove this proposition as an exercise.

Appendix C

Summary of Algebraic Structures Encountered

Loosely speaking, an **algebraic structure** is any set upon which "arithmetic-like" operations have been defined. The importance of such structures in abstract mathematics cannot be overstated. By recognizing a given set S as an instance of a well-known algebraic structure, *every result* that is known about that abstract algebraic structure is then automatically also known to hold for S. This utility is, in large part, the main motivation behind abstraction.

Before reviewing the algebraic structures that are most important to the study of Linear Algebra, we first carefully define what it means for an operation to be "arithmetic-like".

C.1 Binary operations and scaling operations

When we are presented with the *set* of real numbers, say $S = \mathbb{R}$, we expect a great deal of "structure" given on S. E.g., given any two real numbers $r_1, r_2 \in \mathbb{R}$, one can form the sum $r_1 + r_2$, the difference $r_1 - r_2$, the product $r_1 r_2$, the quotient r_1/r_2 (assuming $r_2 \neq 0$), the maximum $\max\{r_1, r_2\}$, the minimum $\min\{r_1, r_2\}$, the average $(r_1 + r_2)/2$, and so on. Each of these operations follows the same pattern: take two real numbers and "combine" (or "compare") them in order to form a new real number. Such operations are called binary operations. In general, a binary operation on an arbitrary non-empty set is defined as follows.

Definition C.1.1. A **binary operation** on a non-empty set S is any function that has as its domain $S \times S$ and as its codomain S.

In other words, a binary operation on S is *any* rule $f : S \times S \to S$ that assigns exactly one element $f(s_1, s_2) \in S$ to each pair of elements $s_1, s_2 \in S$. We illustrate this definition in the following examples.

Example C.1.2.

(1) Addition, subtraction, and multiplication are all examples of familiar binary

operations on $\mathbb{R}$. Formally, one would denote these by something like

$$+ : \mathbb{R} \times \mathbb{R} \to \mathbb{R}, \ - : \mathbb{R} \times \mathbb{R} \to \mathbb{R}, \text{ and } * : \mathbb{R} \times \mathbb{R} \to \mathbb{R}, \text{ respectively.}$$

Then, given two real numbers $r_1, r_2 \in \mathbb{R}$, we would denote their sum by $+(r_1, r_2)$, their difference by $-(r_1, r_2)$, and their product by $*(r_1, r_2)$. (E.g., $+(17, 32) = 49$, $-(17, 32) = -15$, and $*(17, 32) = 544$.) However, this level of notational formality can be rather inconvenient, and so we often resort to writing $+(r_1, r_2)$ as the more familiar expression $r_1 + r_2$, $-(r_1, r_2)$ as $r_1 - r_2$, and $*(r_1, r_2)$ as either $r_1 * r_2$ or $r_1 r_2$.

(2) The division function $\div : \mathbb{R} \times (\mathbb{R} \setminus \{0\}) \to \mathbb{R}$ is *not* a binary operation on $\mathbb{R}$ since it does not have the proper domain. However, division *is* a binary operation on $\mathbb{R} \setminus \{0\}$.

(3) Other binary operations on $\mathbb{R}$ include the maximum function $\max : \mathbb{R} \times \mathbb{R} \to \mathbb{R}$, the minimum function $\min : \mathbb{R} \times \mathbb{R} \to \mathbb{R}$, and the average function $(\cdot + \cdot)/2 : \mathbb{R} \times \mathbb{R} \to \mathbb{R}$.

(4) An example of a binary operation f on the set $S = \{$Alice, Bob, Carol$\}$ is given by

$$f(s_1, s_2) = \begin{cases} s_1 & \text{if } s_1 \text{ alphabetically precedes } s_2, \\ \text{Bob} & \text{otherwise.} \end{cases}$$

This is because *the only requirement* for a binary operation is that exactly one element of S is assigned to every ordered pair of elements $(s_1, s_2) \in S \times S$.

Even though one could define any number of binary operations upon a given non-empty set, we are generally only interested in operations that satisfy additional "arithmetic-like" conditions. In other words, the most interesting binary operations are those that share the salient properties of common binary operations like addition and multiplication on $\mathbb{R}$. We make this precise with the definition of a "group" in Section C.2.

In addition to binary operations defined on pairs of elements in the set S, one can also define operations that involve elements from two different sets. Here is an important example.

Definition C.1.3. A **scaling operation** (a.k.a. **external binary operation**) on a non-empty set S is any function that has as its domain $\mathbb{F} \times S$ and as its codomain S, where $\mathbb{F}$ denotes an arbitrary field. (As usual, you should just think of $\mathbb{F}$ as being either $\mathbb{R}$ or $\mathbb{C}$.)

In other words, a scaling operation on S is *any* rule $f : \mathbb{F} \times S \to S$ that assigns exactly one element $f(\alpha, s) \in S$ to each pair of elements $\alpha \in \mathbb{F}$ and $s \in S$. As such, $f(\alpha, s)$ is often written simply as αs. We illustrate this definition in the following examples.

Example C.1.4.

(1) Scalar multiplication of n-tuples in $\mathbb{R}^n$ is probably the most familiar scaling operation. Formally, scalar multiplication on $\mathbb{R}^n$ is defined as the following function:

$$(\alpha, (x_1, \ldots, x_n)) \longmapsto \alpha(x_1, \ldots, x_n) = (\alpha x_1, \ldots, \alpha x_n), \, \forall\, \alpha \in \mathbb{R}, \, \forall\, (x_1, \ldots, x_n) \in \mathbb{R}^n.$$

In other words, given any $\alpha \in \mathbb{R}$ and any n-tuple $(x_1, \ldots, x_n) \in \mathbb{R}^n$, their scalar multiplication results in a *new n-tuple* denoted by $\alpha(x_1, \ldots, x_n)$. This new n-tuple is virtually identical to the original, each component having just been "rescaled" by α.

(2) Scalar multiplication of continuous functions is another familiar scaling operation. Given any real number $\alpha \in \mathbb{R}$ and any function $f \in \mathcal{C}(\mathbb{R})$, their scalar multiplication results in a *new function* that is denoted by αf, where αf is defined by the rule

$$(\alpha f)(r) = \alpha(f(r)), \forall\, r \in \mathbb{R}.$$

In other words, this new continuous function $\alpha f \in \mathcal{C}(\mathbb{R})$ is virtually identical to the original function f; it just "rescales" the image of each $r \in \mathbb{R}$ under f by α.

(3) The division function $\div : \mathbb{R} \times (\mathbb{R} \setminus \{0\}) \to \mathbb{R}$ is a scaling operation on $\mathbb{R} \setminus \{0\}$. In particular, given two real numbers $r_1, r_2 \in \mathbb{R}$ and any non-zero real number $s \in \mathbb{R} \setminus \{0\}$, we have that $\div(r_1, s) = r_1(1/s)$ and $\div(r_2, s) = r_2(1/s)$, and so $\div(r_1, s)$ and $\div(r_2, s)$ can be viewed as different "scalings" of the multiplicative inverse $1/s$ of s.

This is actually a special case of the previous example. In particular, we can define a function $f \in \mathcal{C}(\mathbb{R} \setminus \{0\})$ by $f(s) = 1/s$, for each $s \in \mathbb{R} \setminus \{0\}$. Then, given any two real numbers $r_1, r_2 \in \mathbb{R}$, the functions $r_1 f$ and $r_2 f$ can be defined by

$$r_1 f(\cdot) = \div(r_1, \cdot) \quad \text{and} \quad r_2 f(\cdot) = \div(r_2, \cdot), \text{ respectively.}$$

(4) Strictly speaking, there is nothing in the definition that precludes S from equalling $\mathbb{F}$. Consequently, addition, subtraction, and multiplication can all be seen as examples of scaling operations on $\mathbb{R}$.

As with binary operations, it is easy to define any number of scaling operations upon a given non-empty set S. However, we are generally only interested in operations that are essentially like scalar multiplication on $\mathbb{R}^n$, and it is also quite common to additionally impose conditions for how scaling operations should interact with any binary operations that might also be defined upon S. We make this precise when we present an alternate formulation of the definition for a vector space in Section C.2.

C.2   Groups, fields, and vector spaces

We begin this section with the following definition, which is one of the most fundamental and ubiquitous algebraic structures in all of mathematics.

Definition C.2.1. Let G be a non-empty set, and let $*$ be a binary operation on
G. (In other words, $* : G \times G \to G$ is a function with $*(a, b)$ denoted by $a * b$, for
each $a, b \in G$.) Then G is said to **form a group** under $*$ if the following three
conditions are satisfied:

(1) (associativity) Given any three elements $a, b, c \in G$,

$$(a * b) * c = a * (b * c).$$

(2) (existence of an identity element) There is an element $e \in G$ such that, given
    any element $a \in G$,

$$a * e = e * a = a.$$

(3) (existence of inverse elements) Given any element $a \in G$, there is an element
    $b \in G$ such that

$$a * b = b * a = e.$$

You should recognize these three conditions (which are sometimes collectively
referred to as the **group axioms**) as properties that are satisfied by the operation of
addition on $\mathbb{R}$. This is not an accident. In particular, given real numbers $\alpha, \beta \in \mathbb{R}$,
the group axioms form the minimal set of assumptions needed in order to solve the
equation $x + \alpha = \beta$ for the variable x, and it is in this sense that the group axioms
are an abstraction of the most fundamental properties of addition of real numbers.

A similar remark holds regarding multiplication on $\mathbb{R} \setminus \{0\}$ and solving the
equation $\alpha x = \beta$ for the variable x. Note, however, that this cannot be extended
to all of $\mathbb{R}$.

The familiar property of addition of real numbers that $a + b = b + a$, is not part
of the group axioms. When it holds in a given group G, the following definition
applies.

Definition C.2.2. Let G be a group under binary operation $*$. Then G is called an
abelian group (a.k.a. **commutative group**) if, given any two elements $a, b \in G$,
$a * b = b * a$.

We now give some of the more important examples of groups that occur in Linear
Algebra, but note that these examples far from exhaust the variety of groups studied
in other branches of mathematics.

Example C.2.3.

(1) If $G \in \{\mathbb{Z}, \mathbb{Q}, \mathbb{R}, \mathbb{C}\}$, then G forms an abelian group under the usual definition of addition.

Note, though, that the set $\mathbb{Z}_+$ of positive integers does not form a group under addition since, e.g., it does not contain an additive identity element.

(2) Similarly, if $G \in \{\mathbb{Q} \setminus \{0\}, \mathbb{R} \setminus \{0\}, \mathbb{C} \setminus \{0\}\}$, then G forms an abelian group under the usual definition of multiplication.

Note, though, that $\mathbb{Z} \setminus \{0\}$ does not form a group under multiplication since only ± 1 have multiplicative inverses.

(3) If $m, n \in \mathbb{Z}_+$ are positive integers and $\mathbb{F}$ denotes either $\mathbb{R}$ or $\mathbb{C}$, then the set $\mathbb{F}^{m \times n}$ of all $m \times n$ matrices forms an abelian group under matrix addition.

Note, though, that $\mathbb{F}^{m \times n}$ does not form a group under matrix multiplication unless $m = n = 1$, in which case $\mathbb{F}^{1 \times 1} = \mathbb{F}$.

(4) Similarly, if $n \in \mathbb{Z}_+$ is a positive integer and $\mathbb{F}$ denotes either $\mathbb{R}$ or $\mathbb{C}$, then the set $GL(n, \mathbb{F})$ of invertible $n \times n$ matrices forms a group under matrix multiplications. This group, which is often called the **general linear group**, is non-abelian when $n \geq 2$.

Note, though, that $GL(n, \mathbb{F})$ does not form a group under matrix addition for any choice of n since, e.g., the zero matrix $0_{n \times n} \notin GL(n, \mathbb{F})$.

In the above examples, you should notice two things. First of all, it is important to specify the operation under which a set might or might not be a group. Second, and perhaps more importantly, all but one example is an abelian group. Most of the important sets in Linear Algebra possess some type of algebraic structure, and abelian groups are the principal building block of virtually every one of these algebraic structures. In particular, fields and vector spaces (as defined below) and rings and algebra (as defined in Section C.3) can all be described as "abelian groups plus additional structure".

Given an abelian group G, adding "additional structure" amounts to imposing one or more additional operations on G such that each new operation is "compatible" with the preexisting binary operation on G. As our first example of this, we add another binary operation to G in order to obtain the definition of a **field**:

Definition C.2.4. Let F be a non-empty set, and let $+$ and $*$ be binary operations on F. Then F **forms a field** under $+$ and $*$ if the following three conditions are satisfied:

(1) F forms an abelian group under $+$.
(2) Denoting the identity element for $+$ by 0, $F \setminus \{0\}$ forms an abelian group under $*$.
(3) ($*$ distributes over $+$) Given any three elements $a, b, c \in F$,

$$a * (b + c) = a * b + a * c.$$

You should recognize these three conditions (which are sometimes collectively referred to as the **field axioms**) as properties that are satisfied when the operations of addition and multiplication are taken together on $\mathbb{R}$. This is not an accident. As with the group axioms, the field axioms form the minimal set of assumptions needed in order to abstract fundamental properties of these familiar arithmetic operations. Specifically, the field axioms guarantee that, given any field F, three conditions are always satisfied:

(1) Given any $a, b \in F$, the equation $x + a = b$ can be solved for the variable x.
(2) Given any $a \in F \setminus \{0\}$ and $b \in F$, the equation $a * x = b$ can be solved for x.
(3) The binary operation $*$ (which is like multiplication on $\mathbb{R}$) can be distributed over (i.e., is "compatible" with) the binary operation $+$ (which is like addition on $\mathbb{R}$).

Example C.2.5. It should be clear that, if $F \in \{\mathbb{Q}, \mathbb{R}, \mathbb{C}\}$, then F forms a field under the usual definitions of addition and multiplication.

Note, though, that the set $\mathbb{Z}$ of integers does not form a field under these operations since $\mathbb{Z} \setminus \{0\}$ fails to form a group under multiplication. Similarly, none of the other sets from Example C.2.3 can be made into a field.

The fields $\mathbb{Q}$, $\mathbb{R}$, and $\mathbb{C}$ are familiar as commonly used number systems. There are many other interesting and useful examples of fields, but those will not be used in this book.

We close this section by introducing a special type of scaling operation called **scalar multiplication**. Recall that $\mathbb{F}$ can be replaced with either $\mathbb{R}$ or $\mathbb{C}$.

Definition C.2.6. Let S be a non-empty set, and let $*$ be a scaling operation on S. (In other words, $* : \mathbb{F} \times S \to S$ is a function with $*(\alpha, s)$ denoted by $\alpha * s$ or even just αs, for every $\alpha \in \mathbb{F}$ and $s \in S$.) Then $*$ is called **scalar multiplication** if it satisfies the following two conditions:

(1) (existence of a multiplicative identity element for $*$) Denote by 1 the multiplicative identity element for $\mathbb{F}$. Then, given any $s \in S$, $1 * s = s$.
(2) (multiplication in $\mathbb{F}$ is quasi-associative with respect to $*$) Given any $\alpha, \beta \in \mathbb{F}$ and any $s \in S$,

$$(\alpha\beta) * s = \alpha * (\beta * s).$$

Note that we choose to have the multiplicative part of $\mathbb{F}$ "act" upon S because we are abstracting scalar multiplication as it is intuitively defined in Example C.1.4 on both $\mathbb{R}^n$ and $\mathcal{C}(\mathbb{R})$. This is because, by also requiring a "compatible" additive structure (called **vector addition**), we obtain the following alternate formulation for the definition of a vector space.

Definition C.2.7. Let V be an abelian group under the binary operation $+$, and let $*$ be a scalar multiplication operation on V with respect to $\mathbb{F}$. Then V **forms**

a vector space over $\mathbb{F}$ with respect to $+$ and $*$ if the following two conditions are satisfied:

(1) ($*$ distributes over $+$) Given any $\alpha \in \mathbb{F}$ and any $u, v \in V$,

$$\alpha * (u + v) = \alpha * u + \alpha * v.$$

(2) ($*$ distributes over addition in $\mathbb{F}$) Given any $\alpha, \beta \in \mathbb{F}$ and any $v \in V$,

$$(\alpha + \beta) * v = \alpha * v + \beta * v.$$

C.3 Rings and algebras

In this section, we briefly mention two other common algebraic structures. Specifically, we first "relax" the definition of a field in order to define a **ring**, and we then combine the definitions of ring and vector space in order to define an **algebra**. Groups, rings, and fields are the most fundamental algebraic structures, with vector spaces and algebras being particularly important within the study of Linear Algebra and its applications.

Definition C.3.1. Let R be a non-empty set, and let $+$ and $*$ be binary operations on R. Then R **forms an (associative) ring** under $+$ and $*$ if the following three conditions are satisfied:

(1) R forms an abelian group under $+$.
(2) ($*$ is associative) Given any three elements $a, b, c \in R$, $a * (b * c) = (a * b) * c$.
(3) ($*$ distributes over $+$) Given any three elements $a, b, c \in R$,

$$a * (b + c) = a * b + a * c \quad \text{and} \quad (a + b) * c = a * c + b * c.$$

As with the definition of group, there are many additional properties that can be added to a ring; here, each additional property makes a ring more field-like in some way.

Definition C.3.2. Let R be a ring under the binary operations $+$ and $*$. Then we call R

- **commutative** if $*$ is a commutative operation; i.e., given any $a, b \in R$, $a * b = b * a$.
- **unital** if there is an identity element for $*$; i.e., if there exists an element $i \in R$ such that, given any $a \in R$, $a * i = i * a = a$.
- a **commutative ring with identity** (a.k.a. **CRI**) if it is both commutative and unital.

In particular, note that a commutative ring with identity is almost a field; the only thing missing is the assumption that every element has a multiplicative inverse. It is this one difference that results in many familiar sets being CRIs (or at least unital rings) but not fields. E.g., $\mathbb{Z}$ is a CRI under the usual operations of addition and multiplication, yet, because of the lack of multiplicative inverses for all elements except ± 1, $\mathbb{Z}$ is not a field.

In some sense, $\mathbb{Z}$ is the prototypical example of a ring, but there are many other familiar examples. E.g., if F is any field, then the set of polynomials $F[z]$ with coefficients from F is a CRI under the usual operations of polynomial addition and multiplication, but again, because of the lack of multiplicative inverses for every element, $F[z]$ is itself not a field. Another important example of a ring comes from Linear Algebra. Given any vector space V, the set $\mathcal{L}(V)$ of all linear maps from V into V is a unital ring under the operations of function addition and composition. However, $\mathcal{L}(V)$ is not a CRI unless $\dim(V) \in \{0, 1\}$.

Alternatively, if a ring R forms a group under $*$ (but not necessarily an abelian group), then R is sometimes called a **skew field** (a.k.a. **division ring**). Note that a skew field is also almost a field; the only thing missing is the assumption that multiplication is commutative. Unlike CRIs, though, there are no simple examples of skew fields that are not also fields.

We close this section by defining the concept of an **algebra over a field**. In essence, an algebra is a vector space together with a "compatible" ring structure. Consequently, anything that can be done with either a ring or a vector space can also be done with an algebra.

Definition C.3.3. Let A be a non-empty set, let $+$ and $\times$ be binary operations on A, and let $*$ be scalar multiplication on A with respect to $\mathbb{F}$. Then A **forms an (associative) algebra** over $\mathbb{F}$ with respect to $+$, $\times$, and $*$ if the following three conditions are satisfied:

(1) A forms an (associative) ring under $+$ and $\times$.
(2) A forms a vector space over $\mathbb{F}$ with respect to $+$ and $*$.
(3) ($*$ is quasi-associative and homogeneous with respect to $\times$) Given any element $\alpha \in \mathbb{F}$ and any two elements $a, b \in R$,

$$\alpha * (a \times b) = (\alpha * a) \times b \quad \text{and} \quad \alpha * (a \times b) = a \times (\alpha * b).$$

Two particularly important examples of algebras were already defined above: $F[z]$ (which is unital and commutative) and $\mathcal{L}(V)$ (which is, in general, just unital). On the other hand, there are also many important sets in Linear Algebra that are not algebras. E.g., $\mathbb{Z}$ is a ring that cannot easily be made into an algebra, and $\mathbb{R}^3$ is a vector space but cannot easily be made into a ring (note that the cross product operation from Vector Calculus is not associative).

Appendix D

Some Common Mathematical Symbols
and Abbreviations
(with History)

This Appendix contains a list of common mathematical symbols as well as a list of common Latin abbreviations and phrases. While you will not necessarily need all of the included symbols for your study of Linear Algebra, this list will give you an idea of where much of our modern mathematical notation comes from.

Binary Relations

$=$ (the **equals sign**) means "is the same as" and was first introduced in the 1557 book *The Whetstone of Witte* by physician and mathematician Robert Recorde (c. 1510–1558). He wrote, "I will sette as I doe often in woorke use, a paire of parralles, or Gemowe lines of one lengthe, thus: ======, bicause noe 2 thynges can be moare equalle." (Recorde's equals sign was significantly longer than the one in modern usage and is based upon the idea of "Gemowe" or "identical" lines, where "Gemowe" means "twin" and comes from the same root as the name of the constellation "Gemini".)

Robert Recorde also introduced the **plus sign**, "$+$", and the **minus sign**, "$-$", in *The Whetstone of Witte*.

$<$ (the **less than sign**) means "is strictly less than", and $>$ (the **greater than sign**) means "is strictly greater than". These first appeared in the book *Artis Analyticae Praxis ad Aequationes Algebraicas Resolvendas* ("The Analytical Arts Applied to Solving Algebraic Equations") by mathematician and astronomer Thomas Harriot (1560–1621), which was published posthumously in 1631.

Pierre Bouguer (1698–1758) later refined these to $\leq$ ("is less than or equals") and $\geq$ ("is greater than or equals") in 1734. Bouger is sometimes called "the father of naval architecture" due to his foundational work in the theory of naval navigation.

$:=$ (the **equal by definition sign**) means "is equal by definition to". This is a common alternate form of the symbol "$=_{\text{Def}}$", the latter having first appeared in the 1894 book *Logica Matematica* by logician Cesare Burali-Forti (1861–1931).

Other common alternate forms of the symbol "$=_{\text{Def}}$" include "$\overset{\text{def}}{=}$" and "$\equiv$", with "$\equiv$" being especially common in applied mathematics.

$\approx$ (the **approximately equals sign**) means "is approximately equal to" and was first introduced in the 1892 book *Applications of Elliptic Functions* by mathematician Alfred Greenhill (1847–1927).

Other modern symbols for "approximately equals" include "$\doteq$" (read as "is nearly equal to"), "$\cong$" (read as "is congruent to"), "$\simeq$" (read as "is similar to"), "$\asymp$" (read as "is asymptotically equal to"), and "$\propto$" (read as "is proportional to"). Usage varies, and these are sometimes used to denote varying degrees of "approximate equality" within a given context.

Some Symbols from Mathematical Logic

$\therefore$ (**three dots**) means "therefore" and first appeared in print in the 1659 book *Teusche Algebra* ("Teach Yourself Algebra") by mathematician Johann Rahn (1622–1676).

Teusche Algebra also contains the first use of the **obelus**, "$\div$", to denote division.

$\because$ (**upside-down dots**) means "because" and seems to have first appeared in the 1805 book *The Gentleman's Mathematical Companion*. However, it is much more common (and less ambiguous) to just abbreviate "because" as "b/c".

$\ni$ (the **such that sign**) means "under the condition that" and first appeared in the 1906 edition of *Formulaire de mathématiques* by the logician Giuseppe Peano (1858–1932). However, it is much more common (and less ambiguous) to just abbreviate "such that" as "s.t.".

There are two good reasons to avoid using "$\ni$" in place of "such that". First of all, the abbreviation "s.t." is significantly more suggestive of its meaning than is "$\ni$". More importantly, the symbol "$\ni$" is now commonly used to mean "contains as an element", which is a logical extension of the usage of the standard symbol "$\in$" to mean "is contained as an element in".

$\Rightarrow$ (the **implies sign**) means "logically implies that", and $\Leftarrow$ (the **is implied by sign**) means "is logically implied by". Both have an unclear historical origin. (E.g., "if it's raining, then it's pouring" is equivalent to saying "it's raining $\Rightarrow$ it's pouring.")

$\Longleftrightarrow$ (the **iff** symbol) means "if and only if" (abbreviated "iff") and is used to connect two logically equivalent mathematical statements. (E.g., "it's raining iff it's pouring" means simultaneously that "if it's raining, then it's pouring" and that "if it's pouring, then it's raining". In other words, the statement "it's raining $\Longleftrightarrow$ it's pouring" means simultaneously that "it's raining $\Rightarrow$ it's pouring" and "it's raining $\Leftarrow$ it's pouring".)

The abbreviation "iff" is attributed to the mathematician Paul Halmos

(1916–2006).

∀ (the **universal quantifier**) means "for all" and was first used in the 1935 publication *Untersuchungen über das logische Schliessen* ("Investigations on Logical Reasoning") by logician Gerhard Gentzen (1909–1945). He called it the *All-Zeichen* ("all character") by analogy to the symbol "∃", which means "there exists".

∃ (the **existential quantifier**) means "there exists" and was first used in the 1897 edition of *Formulaire de mathématiques* by the logician Giuseppe Peano (1858–1932).

□ (the **Halmos tombstone** or **Halmos symbol**) means "Q.E.D.", which is an abbreviation of the Latin phrase *quod erat demonstrandum* ("which was to be proven"). "Q.E.D." has been the most common way to symbolize the end of a logical argument for many centuries, but the modern convention of the "tombstone" is now generally preferred both because it is easier to write and because it is visually more compact.

The symbol "□" was first made popular by mathematician Paul Halmos (1916–2006).

Some Notation from Set Theory

⊂ (the **is included in sign**) means "is a subset of" and ⊃ (the **includes sign**) means "has as a subset". Both symbols were introduced in the 1890 book *Vorlesungen über die Algebra der Logik* ("Lectures on the Algebra of the Logic") by logician Ernst Schröder (1841–1902).

∈ (the **is in sign**) means "is an element of" and first appeared in the 1895 edition of *Formulaire de mathématiques* by the logician Giuseppe Peano (1858–1932). Peano originally used the Greek letter "ϵ" (viz. the first letter of the Latin word *est* for "is"). The modern stylized version of this symbol was later introduced in the 1903 book *Principles of Mathematics* by logician and philosopher Betrand Russell (1872–1970).

It is also common to use the symbol "∋" to mean "contains as an element", which is not to be confused with the more archaic usage of "∋" to mean "such that".

∪ (the **union sign**) means "take the elements that are in either set", and ∩ (the **intersection sign**) means "take the elements that the two sets have in common". These were both introduced in the 1888 book *Calcolo geometrico secondo l'Ausdehnungslehre di H. Grassmann preceduto dalle operazioni della logica deduttiva* ("Geometric Calculus based upon the teachings of H. Grassman, preceded by the operations of deductive logic") by logician Giuseppe Peano (1858–1932).

∅ (the **null set** or **empty set**) means "the set without any elements in it" and was first used in the 1939 book *Éléments de mathématique* by Nicolas Bourbaki. (Bourbaki is the collective pseudonym for a group of primarily European mathematicians who have written many mathematics books together.) It was borrowed simultaneously from the Norwegian, Danish and Faroese alphabets by group member André Weil (1906–1998).

∞ (**infinity**) denotes "a quantity or number of arbitrarily large magnitude" and first appeared in print in the 1655 publication *De Sectionibus Conicus* ("Tract on Conic Sections") by mathematician John Wallis (1616–1703).

Possible explanations for Wallis' choice of "∞" include its resemblance to the symbol "*oo*" (used by ancient Romans to denote the number 1000), to the final letter of the Greek alphabet ω (used symbolically to mean the "final" number), and to a simple curve called a "lemniscate", which can be endlessly traversed with little effort.

Some Important Numbers in Mathematics

π (the **ratio of the circumference to the diameter of a circle**) denotes the number $3.141592653589\ldots$, and was first used in the 1706 book *Synopsis palmariorum mathesios* ("A New Introduction to Mathematics") by mathematician William Jones (1675–1749). The use of π to denote this number was then popularized by the great mathematician Leonhard Euler (1707–1783) in his 1748 book *Introductio in Analysin Infinitorum*. (It is speculated that Jones chose the letter "π" because it is the first letter in the Greek word *perimetron*, $\pi\epsilon\rho\iota\mu\epsilon\tau\rho\rho\nu$, which roughly means "around".)

$e = \lim_{n\to\infty}(1 + \frac{1}{n})^n$ (the **natural logarithm base**, also sometimes called **Euler's number**) denotes the number $2.718281828459\ldots$, and was first used in the 1728 manuscript *Meditatio in Experimenta explosione tormentorum nuper instituta* ("Meditation on experiments made recently on the firing of cannon") by Leonhard Euler. (It is speculated that Euler chose "e" because it is the first letter in the Latin word for "exponential".)

The mathematician Edmund Landau (1877–1938) once wrote that, "The letter e may now no longer be used to denote anything other than this positive universal constant."

$i = \sqrt{-1}$ (the **imaginary unit**) was first used in the 1777 memoir *Institutionum calculi integralis* ("Foundations of Integral Calculus") by Leonhard Euler.

The five most important numbers in mathematics are widely considered to be (in order) 0, 1, i, π, and e. These numbers are even remarkably linked by the equation $e^{i\pi} + 1 = 0$, which the physicist Richard Feynman (1918–1988) once called "the most remarkable formula in mathematics".

$\gamma = \lim_{n\to\infty}(\sum_{k=1}^{n} \frac{1}{k} - \ln n)$ (the **Euler-Mascheroni constant**, also known as just **Euler's constant**), denotes the number $0.577215664901\ldots$, and was first used in the 1792 book *Adnotationes ad Euleri Calculum Integralem* ("Annotations to Euler's Integral Calculus") by geometer Lorenzo Mascheroni (1750–

1800).

The number γ is widely considered to be the sixth most important number in mathematics due to its frequent appearance in formulas from number theory and applied mathematics. However, as of this writing, it is still not even known whether or not γ is an irrational number.

Some Common Latin Abbreviations and Phrases

i.e. (*id est*) means "that is" or "in other words". (It is used to paraphrase a statement that was just made, **not** to mean "for example", and is **always** followed by a comma.)

e.g. (*exempli gratia*) means "for example". (It is usually used to give an example of a statement that was just made and is **always** followed by a comma.)

viz. (*videlicet*) means "namely" or "more specifically". (It is used to clarify a statement that was just made by providing more information and is **never** followed by a comma.)

etc. (*et cetera*) means "and so forth" or "and so on". (It is used to suggest that the reader should infer further examples from a list that has already been started and is **usually not** followed by a comma.)

et al. (*et alii*) means "and others". (It is used in place of listing multiple authors past the first. The abbreviation "et al." can also be used in place of *et alibi*, which means "and elsewhere".)

cf. (*conferre*) means "compare to" or "see also". (It is used either to draw a comparison or to refer the reader to somewhere else that they can find more information, and it is **never** followed by a comma.)

q.v. (*quod vide*) means "which see" or "go look it up if you're interested". (It is used to cross-reference a different written work or a different part of the same written work, and it is **never** followed by a comma.) The plural form of "q.v." is "q.q."

v.s. (*vide supra*) means "see above". (It is used to imply that more information can be found before the current point in a written work and is **never** followed by a comma.)

N.B. (*Nota Bene*) means "note well" or "pay attention to the following". (It is used to imply that the wise reader will pay especially careful attention to what follows and is **never** followed by a comma. Cf. the abbreviation "verb. sap.")

verb. sap. (*verbum sapienti sat est*) means "a word to the wise is enough" or "enough has already been said". (It is used to imply that, while something may still be left unsaid, enough has been said for the reader to infer the entire meaning.)

vs. (*versus*) means "against" or "in contrast to". (It is used to contrast two things and is **never** followed by a comma.) The abbreviation "vs." is also often written as "v."

   c. (*circa*) means "around" or "near". (It is used when giving an approximation, usually for a date, and is **never** followed by a comma.) The abbreviation "c." is also commonly written as "ca.", "cir.", or "circ."

ex lib. (*ex libris*) means "from the library of". (It is used to indicate ownership of a book and is **never** followed by a comma.).

- *vice versa* means "the other way around" and is used to indicate that an implication can logically be reversed. (This is sometimes abbreviated as "v.v.")

- *a fortiori* means "from the stronger" or "more importantly".

- *a priori* means "from before the fact" and refers to reasoning that is done while an event still has yet to happen.

- *a posteriori* means "from after the fact" and refers to reasoning that is done after an event has already happened.

- *ad hoc* means "to this" and refers to reasoning that is specific to an event as it is happening. (Such reasoning is regarded as not being generalizable to other situations.)

- *ad infinitum* means "to infinity" or "without limit".

- *ad nauseam* means "causing sea-sickness" or "to excess".

- *mutatis mutandis* means "changing what needs changing" or "with the necessary changes having been made".

- *non sequitur* means "it does not follow" and refers to something that is out of place in a logical argument. (This is sometimes abbreviated as "non seq.")

Index

Printed in the United States
By Bookmasters